YOUNGER FOR LIFE

YOUNGER
FOR
LIFE

FEEL GREAT AND LOOK YOUR BEST
WITH THE NEW SCIENCE OF
AUTOJUVENATION

ANTHONY YOUN, M.D.

HANOVER
SQUARE
PRESS

HANOVER
SQUARE
PRESS™

ISBN-13: 978-1-335-00787-2

Younger for Life

Hanover Square Press
22 Adelaide St. West, 41st Floor
Toronto, Ontario M5H 4E3, Canada
HanoverSqPress.com
BookClubbish.com

Printed in U.S.A.

For Amy, the only person I want to grow old with.

Also by Anthony Youn, M.D.

Playing God: The Evolution of a Modern Surgeon

The Age Fix: A Leading Plastic Surgeon Reveals How to Really Look 10 Years Younger

In Stitches

YOUNGER
FOR
LIFE

This book is written to provide information on many of the various options for health and beauty that are currently available as of this publication date. It is based solely on the opinions of Dr. Anthony Youn, whose opinions may not reflect every doctor's standpoint. Before you undergo any cosmetic procedure, begin a skin care regimen or make major dietary changes, it is important that you consult with your physician to ensure that it won't adversely affect your health. Following any of the recommendations in this book does not constitute a doctor-patient relationship, and the author and publisher expressly disclaim any responsibility for any adverse effects arising from the use or application of the information contained herein.

TABLE OF CONTENTS

Part Five: Revitalize With An Anti-Aging Lifestyle

Part Six: Regenerate With Next-Level Holistic Anti-Aging Treatments

TESTIMONIALS

FROM REAL PEOPLE WHO TRIED THE YOUNGER FOR LIFE AUTOJUVENATION JUMP START

"I am honestly *amazed*. I can't believe what an all-over benefit I got from this. I don't look in the mirror and cringe anymore. I also *loved* that I didn't feel restricted in any way. The whole family ate with me and loved it!"

—B. C., age 53

"My face is so *smooth* and feels hydrated pretty much all day, which is something I was struggling with. I have seen huge changes in my chest and hands, too!"

—A. R., age 54

"My friends have noticed the improvement in my skin—looking younger, glowing, and *fresh*. They're surprised I'm not wearing makeup and look like this with a fresh face!"

—S. A., age 44

"After completing the three-week Autojuvenation Jump Start, my face appears fuller, the dark circles around my eyes are less prominent, I have more color in my face, and there's a reduction in small wrinkles. My husband says my skin looks 37% better! Within eight days of starting the program, I had two unsolicited comments about how great my face looks!"

—L. F., age 59

"I've had positive comments that I look great and my face is apparently glowing! This program helped get my mind and body into healthy habits."

—P. B., age 56

"My skin is smoother, I feel energized, and I know that I'm doing something good for my health and appearance!"

—D. A., age 46

INTRODUCTION:

HOW OLD ARE YOU, REALLY?

I'm a Boomer.

Well, not really in the classic sense, as Baby Boomers were traditionally born between 1946 and 1964. Technically I'm Generation X, yet young people, and especially my millions of fans on social media, like to call me a Boomer because to *them* I seem old. In their eyes, *Boomer* is a word *young* people use to describe *old* people—people who are old and therefore surely clueless about life in today's modern world.

I don't like to be insulted on my own social media accounts, so I've blocked any comment that includes the term "boomer." I've also blocked the terms "simp," "shart," and "fivehead," but those are whole other matters. The main thing is that I don't like being reminded of my age, and the fact that I'm no longer considered the "young doctor." But notice how I emphasize the words *old* and *young* with italics. That's because the meaning of these terms has evolved.

Because you've picked up this book, I assume you know what

it feels like to be what you think of as old. Whether in looks, attitude, or energy level, whether you are pushing 30 or 50 or even 90, something has changed.

Maybe you're one of my TikTok followers in your mid-to-late-twenties and are just now noticing that your skin doesn't bounce back the way it used to after an indulgent night out. You could be in your thirties and seeing *11* lines, crow's feet, and frown lines, or more than a few gray hairs, and you wonder if you can reverse what's happening. Maybe you're like me, straddling your forties and fifties, and noticing the skin under your neck starting to sag, right along with your former level of vim and vigor. Or quite possibly you are in your sixties and seventies and seeking ways to keep looking as young as you feel inside—or feeling as young as you used to feel.

Whoever you are, whatever your age, if aging concerns have casually crossed your mind or are constantly weighing on you, this book has solutions.

Although I believe that aging is a blessing (the alternative is certainly undesirable), it's no fun to look in the mirror and see a face that doesn't project the youthful vibrancy you would like. It can be disheartening to notice that you're slowing down or don't feel like yourself anymore on the inside.

But can you really do something about aging? Can you really rejuvenate, both inside and out? Can you reverse the course of slow decline? Of course not.

Or...can you?

I believe that, in many ways, yes, you can look and feel younger for life by halting or even reversing the processes that cause the signs of aging we all know too well. I call this process *autojuvenation* and it is what this book is all about—I'm going to show you simple methods for encouraging your body to rejuvenate itself, so that rejuvenation becomes almost automatic. When you follow the Younger for Life Program, you can harness your body's own natural rejuvenative powers in ways that can help you to look and feel younger.

The fact that you've picked up this book means that you're not just going to let aging happen to you. You have control over how fast it goes, and with this book, you'll be able to autojuvenate yourself back to a younger you.

I don't mean literally turning back the clock, of course. I'll never be 20 or 30 again and frankly wouldn't want to go through the trials and tribulations of those less secure years. (Okay, maybe I'd take a time-travel trip back to age 30...) Once you've passed those landmarks, you can't go back either. However, if much of aging is about the damage caused by environment and lifestyle (and it is), we absolutely can slow it or even reverse the processes that are causing it to happen before it really must. We can go back in time in many ways: in how we feel, look, and act, while still retaining the wisdom of our years and our better, healthier habits (which you'll learn about in this book).

Reversing the premature aging caused by lifestyle doesn't require a souped-up DeLorean time machine, a special pill, or even plastic surgery (and I say this as someone who makes his living doing plastic surgery). All it takes is you doing a few things a little differently.

Humans are meant to age. Aging is inevitable as long as time is inevitable. However, we are not meant to age the way most of us are currently aging. I think we've forgotten what natural aging even looks like. We aren't meant to decline slowly and suffer with pain, disease, and disability for years before our time is up. Aging isn't supposed to feel bad. That is neither normal nor inevitable.

You can control the major factors that influence premature, unnatural aging in most people. If you know how to eat, take care of your skin, provide your body with the right nutrients, manage your stress, and have a young attitude, you really can slow, stop, or even reverse many of those processes that keep you from staying healthy and functional into old age. I'm talking about inflammation, collagen degradation, oxidative stress from exposure to sun and pollution and harmful chemicals, chronic stress, suboptimal nutrition, and what I call *old thinking*.

In this book, I'm going to show you:

- how to be choosier about the quality and purity of your food, as well as when and how much you eat;
- how to take care of your skin differently so it can bounce back to how it looked years ago; and
- how to learn to embrace the lifestyle, attitude, motivation, and habits of a much younger you.

Trust me when I tell you that it is *never* too late to reverse the aging process. Wherever you are now, you can turn back the clock. And if you are just seeing those pesky signs of aging rearing their obnoxious little heads, then it's possible for you to drastically slow down their progression. Let's kick aging to the curb!

Because I treat people every day who are worried about aging, I hear what their concerns are, and I know what people are looking for as they seek youth and beauty. They want to look younger, but what they really want is to *be younger*. So why get a facelift when you could get a *life lift*?

I predict you're going to love what happens to your face, not to mention the rest of you, when you change how you eat, switch to a clean skin care routine, upgrade your sleep, channel your stress, experiment with age-defying intermittent fasting, spend some time doing yoga and meditation (really, it's not difficult), and start thinking like the person you were before you started feeling old.

These are just some of the cures you'll find within these pages, and I know they work because I have tested them on my patients and on myself. You'll read some of the testimonials from my patients and followers, and you'll see how effective the Younger for Life Program can be—and best of all, it doesn't require plastic surgery.

In this book, I'll introduce you to the two-phase Younger for Life Diet. Phase 1 is about nourishing and rebuilding collagen, and Phase 2 is about increasing cellular rejuvenation.

Next, I'll give you a transformative clean-beauty regimen you can use for the rest of your life.

I'll put the two together in the form of an easy-to-follow three-week Autojuvenation Jump Start.

Then I'll cover the most important lifestyle changes for reversing the aging process at the cellular level, including instructions for sleep hygiene, yoga, exercise, meditation, and natural dental care.

Finally, for those who are interested, I'll tell you about some of the more advanced (but noninvasive or only minimally invasive) technologies that can target the damage caused by lifestyle and environment, including everything you ever wanted to know about Botox, fillers, microneedling, chemical peels, red light, fat-blasting treatments, and fractional lasers, so you can decide for yourself if any of these are right for you. Maybe they are, but with an anti-aging diet and lifestyle, you may decide that you don't need them because you are so happy with the results from simple lifestyle changes.

With Younger for Life, you will be aging in reverse. You'll see *and feel* the difference. You might even suspect that you really *have* turned back the clock 10 or even 20 or more years. Maybe this book is the time machine you've been waiting for (minus the DeLorean).

DIDN'T I ALREADY FIX THIS PROBLEM?

In my last book, *The Age Fix*, my premise was that external signs of aging can be fixed, mostly through noninvasive or minimally invasive techniques and procedures. I wanted to help people achieve an appearance that matched the youthful way they felt on the inside. But that book was aptly named because it was indeed about fixes—ways to *look* and *seem* younger. I was happy with that book, even though some readers didn't like that I included information about actual plastic surgeries. (I am a plastic surgeon, after all.) But what I didn't talk about enough (in retrospect) was the part about how old people *feel*.

I've been pondering the difference between the appearance of aging and the feeling of aging for several years now, but I've only recently come to a point in my life and career where I could really search for the answers. I've been busy! That's because after *The Age Fix* came out, my practice hit a pinnacle of success—or so I thought.

There is an idea, in the world of medicine, that the bigger the surgery you perform, the bigger the prestige and the more successful you are. When surgeons are residents in training, we do lower-level operations, hoping that someday we'll be experienced enough to do the big ones. For general surgeons, the ability to perform a Whipple procedure is considered the pinnacle of surgical prowess. This is a complicated and difficult eight-to-ten-hour cancer surgery. And no, you are not allowed to go to the bathroom in the middle of it. (People really do ask that.)

For plastic surgeons, our Whipple procedure is the facelift. Facelifts are complex and difficult, and they cost the most of any plastic surgery. Heck, during the operation, we are literally peeling the skin off a person's face! People may not care too much who does their liposuction, but they are very picky, and rightly so, about who does their facelift.

When I started out in residency, I spent a lot of time pulling out unwanted varicose veins. A year later, I was scrubbing in on skin-cancer removals, and after that, trauma reconstruction. Most plastic surgery residents only begin assisting and performing facelifts when they hit their final year of residency, when they are almost done with their training and are bona fide surgeons themselves.

So once I was in practice, I gauged my success by how many facelifts I had on my schedule. Lots of facelifts meant lots of success. After *The Age Fix* came out, I had a one-year waiting list for facelifts. Then one afternoon, I had a patient who experienced a horrible complication from a facelift I performed. I was devastated. I went over and over in my mind how I might have prevented the tragedy. While I don't think there was anything I could have done differently, it was a terrible, heartbreaking event, and I began to ask myself: What if the ultimate goal of a plastic surgeon isn't about doing the most complex, expensive, or prestigious surgeries? What if it's just the opposite? What if the real goal, the true sign of greatness, was to keep people out of the operating room altogether?

It's not exactly a logical thought. How can a surgeon who doesn't perform surgeries be successful? But I began to look at my profession and how I practice medicine in a new light after that event. There will always be those who want facelifts and will only get the changes they are looking for by going under the knife. But there are many, many more people who would never get a facelift—people who would still very much like to get the *effect* of one—who want to look dramatically younger on the outside but without the trauma and cost of having actual surgery. Wouldn't it mean more if I could help people to look, feel, and actually *be* younger, without ever booking a single plastic surgery procedure?

In plastic surgery, we do many different kinds of so-called anti-aging procedures, but they don't actually reverse aging. These procedures may make you look younger, but the cells in your body won't function in the way younger cells do. I know plenty of people with 30- and 40-year-old faces who have 50- and 60-year-old insides, including celebrities and influencers. This distinction began to interest me more and more. How could I fix *that problem*? How might I impact people who are aging internally more quickly than is natural?

This new ambition led me into the realm of holistic medicine, where the focus is not so much on the symptoms but the root cause of dysfunction. Instead of trying to erase wrinkles and lift drooping skin, or accepting low energy, joint pain, poor sleep, unwanted weight gain, and brain fog as "just aging," this approach to health focuses on the *why*. Why are you developing wrinkles, sagging, fatigue, degraded posture, chronic pain, and that general, difficult-to-define state of just feeling *old*? Because your cells are aging faster than is necessary. Once you realize this, you can begin to change the way your body is aging at the cellular level by changing how you live.

In holistic medicine, that is where the treatment starts, and in this book, that's where autojuvenation starts: at the cause end, not at the symptoms end. Holistic and forward-thinking doctors look at why joints are getting stiff and creaky, why digestion isn't working as well, why cellular turnover has slowed, why hair fades and thins, why skin wrinkles and crepes, and why arteries harden and insulin stops working.

When I began to look to the root causes of aging, it changed my practice permanently. I'm still a plastic surgeon. I'm not a nutritionist, fitness trainer, endocrinologist, cardiologist, or neurologist, and I won't ever pretend to be any of those things. I still do surgeries for those who want them, but I have also spent thousands of hours over the last ten years learning about a way of treating patients that I was never taught in medical school or in residency.

I've learned that food is medicine, that meditation adjusts brain chemistry and structure, that movement triggers changes in the body that don't happen in sedentary people, and that what you put *in* your body, as well as what you put *on* it, can transform how your skin ages. Bottom line: How you live will change how you age, and how you age influences how old you feel and how well you will function *as well as* how you look. In short, how you live can accelerate aging or it can autojuvenate you.

That said, aging is most certainly not a disease. Thinking of aging this way is a pretty recent cultural development. There is so much you can do that will help you to look and feel better that you will never hear about from your regular family doctor (or dermatologist, gynecologist, or endocrinologist). But you're going to hear it from me, and I think you're going to like what you hear. Plastic surgery is a fix, but we can do more than patch up symptoms. The key is autojuvenation, so we can be younger for life.

PART ONE:

THE CAUSES OF AGING

CHAPTER ONE:

WHAT IS AGING, AND HOW GRACEFULLY ARE YOU DOING IT?

"How old are you?"

Most people have the common sense (or courtesy) not to ask this question of just anybody, but I get asked this all the time. When I started my practice in 2004 at the tender age of 31, patients would ask me this because they thought I looked too young to be a plastic surgeon. Unfortunately, those times are long gone, and now I'm asked about my age as more of a curiosity than anything. I once had a patient coming out of anesthesia after an operation ask the nurse, "How does Dr. Youn have so much hair when he's *so old*?"

On social media, all the gloves are off in my comments section.

"This doctor looks like he's 20, but I think he's 60."

"You're as old as my grandma but could pass as my cousin."

"Why am I attracted to a man older than my dad? Yuck! And yum!"

Or my favorite, "His hair is black on his head, but I bet he's completely gray *down there*."

After almost 20 years in private practice, I empathize with my patients when they stand in front of the mirror and lift up the sides of their face to get a wistful glimpse of what they used to look like when they were younger. In our youth-obsessed culture, it's totally natural to long for the look of a younger you and to want to reverse the signs of aging. People want to know if I practice what I preach or if my suggestions work, based on how old *I* look for *my* age. That's why I so often get this question.

If you google me, you'll find out my exact age and birthday. (It's pretty creepy that the internet has this information completely correct and anyone can access it.) Still, I have fun answering this question when it's posed to me online. I amuse myself by replying with random ages, ranging from 21 to 85. And the crazy thing is that people believe me, no matter what I answer!

"Really? You look great for 85!"

"Wow. I didn't know you were 24. I thought you were 40! You look like crap! Use your own skin care!"

"That's cool that you were a fighter pilot in World War II!"

Jokes and fantasy ages aside, aging is something most people begin to think about at some point, some earlier than others. I have patients in their late twenties already worried about wrinkles. Despite the old saying that age is just a number, aging is much more complicated than either the number of years since your birth or that elusive *you're only as old as you feel*. What if you *feel* old but you want to feel young? What if your looks and how you feel don't match? What if you want to be young, not just on the surface but inside as well, with more energy, more vitality, a quicker mind, stronger muscles, and a heart, lungs, liver, kidneys, and hormone balance of someone in their prime? Many people want to know: Is that possible? Could it be possible? *Dr. Youn, please, tell us this is possible!*

Aging is a funny thing (or not so funny, when it seems to be happening to you in a fast and furious manner). One day, you feel pretty much like yourself, and the next day, you wake up and it's a little harder to get out of bed. You're feeling strangely *not* like yourself, and then you look in the mirror, and holy cow... "How did I go from being a Spice Girl to being a Golden Girl?" Or for me, "When did I get my dad's gray hair and permafrown?" At times like these, you may find yourself dreaming about how great it would be if there was a cure for aging.

Are you hoping I'm going to tell you there is?

I am! But before we get into the nuts and bolts of reversing the signs of aging that can be so detrimental to living the life you want to be living right now, let's talk about what aging really is…and isn't.

A BRIEF HISTORY (AND THE MYTHOLOGY) OF AGING

Aging is hardly a concern exclusive to the twenty-first century. Humans have been fascinated and obsessed by the idea of aging probably since they could see their own reflections in the surface of the water or feel their own bodies slowing down. Our preoccupation with aging is as old as the most ancient civilizations. Maybe it's the universally human fear of death, but I think it's more than that. We all know what it feels like to be young and vigorous, and I think you notice when you no longer feel that way. You want that energy back. You want to reclaim that luminous beauty that all young people have (whether they see it at the time or not).

I know I feel very different about my age now than I did back in my twenties. A 2018 study[1] asked people between the ages of 10 and 89 questions about aging, and found that the older people were, the more they wanted to live longer. Older people were also more likely to say they felt younger (mentally) than they really were. Younger people, on the other hand, were generally less concerned with longevity. Of course, so-called old age isn't yet looming when you're in your twenties. They were also less likely to say they felt younger than their age.

Attitudes about aging have certainly changed. Two hundred years ago, older people were more often honored, celebrated, and

29

respected. This is still the case in certain countries, such as my family's country of origin, South Korea. In fact, every January 1, our tradition is to bow to our elders, and in exchange they give us money. Now, to be totally honest, as a kid, I didn't mind the practice, as visions of my grandma dropping me a Benjamin Franklin danced before my eyes. Unfortunately, once I checked out the folded bill she'd slipped into my palm, I usually found myself face-to-face with George Washington instead. Oh well, a kid can dream, right?

We have nothing like this tradition here in the States, although my children probably wouldn't mind bowing to my greatness as long as I slipped them at least a ten-spot. So maybe it's always been the money more than valuing the wisdom of our elders. Or the fear: in some cultures, the belief that generations of ancestors are constantly watching and judging us keeps some people respectful, not just of their older family members but of those who have long passed on to the other side.

But that respect has seemingly faded, especially in the Western world. Throughout the twentieth century, as aging has become increasingly medicalized, older adults are more often portrayed as incompetent, ill, or just cute or hilarious—and, ultimately dismissed, or at least not taken very seriously.

Considering our youth-centric culture, of course we don't want to think of ourselves as old or sick or silly. The truth is that age has many benefits, including wisdom, experience, and perspective, but it's hard to focus on that when we all get the message that younger is better. For better or for worse, the West now largely influences other cultures. I fear that traditions such as bowing to your elders will soon be forgotten everywhere.

Bowing notwithstanding, people have been trying to stave off aging—or at least the look of it—since the dawn of civilization. There are records from ancient Egypt of wrinkle creams. In ancient India, aging was considered the result of internal disharmony, and in ancient China, signs of aging were treated with herbal and dietary remedies to restore internal balance.*

Some cultures went so far as to say that life really could be excessively prolonged. The ancient yogis believed that mastering prana, or life-force energy, was the secret to the cessation of the aging process, and mythologically, some yogis have claimed to live for hundreds of years, often through extreme practices, like (allegedly) not eating anything but air and sunlight.[2] Even if it was possible to survive this long via these practices, who would want to live like that? Definitely not me!

Most of us are probably aware of the centuries-old people in the Old Testament of the Bible. Adam was said to have died at 930 years old, and Noah supposedly reached 950. Methuselah beat them both, living to 969 years! In ancient Buddhist lore, some wise sages (especially Buddhas) supposedly lived for 100,000 years or more.[3] If you look at longevity claims from various Asian cultures as well as ancient Greece and ancient Egypt, you can find reports (or at least tall tales) of people living for tens of thousands, hundreds of thousands, or in some cases, millions of years.[4] They may not have been fresh-faced when they finally crossed over, but if you believe the stories (and that's a stretch), these oldsters sure managed to cheat death for quite a while.

AGING IRL

These concerns—and ambitions—are still with us. My friend and author of the *New York Times* bestseller *Superhuman*, Dave Asprey, has made it his goal to live to 180 using a combination of diet, fasting, and biohacking. Given the tales from long ago, it's at least possible that he could reach or come close to this goal, and he's not the only biohacker with extreme longevity aspirations.

The more we learn about aging and how it works in the

---

* Article by Dr. Mark E. Williams, who specializes in geriatric medicine and is the author of *The Art and Science of Aging Well*

human body (see the next chapter), the more we really might be able to intervene significantly in the aging process—or at least its negative side effects—until the very end. We may also be on the verge of understanding how to prolong that end to the biologically possible maximum, whatever that may turn out to be. The oldest person ever to have lived, whose age has been officially verified, was a French woman named Jeanne Calment,[5] who lived to be 122 years and 164 days old. She died in 1997. Considering how much more we know now about health and wellness than we did in 1997, is it really too much to expect that we could add another 50+ years to that record?

Of course, there is a difference between trying to cure aging and trying to avoid death altogether. I'm pretty sure we haven't figured out how to outsmart the Grim Reaper just yet. He's coming for us one way or another, but in the meantime, what can we do about how we feel, how functional we are, and okay, yes, how we look, in that space between adulthood and The End?

Many people would agree that living for thousands of years isn't worthwhile if you *feel* thousands of years old. We all know of or have heard of people living very difficult existences in the latter years of their lives, basically being propped up and kept alive by modern medical care. This is not how I want to age, and I assume you agree. But I'd like to push 100 if I can (or join Dave at 180!), while still feeling spry, being mobile, and having my wits about me. I'd like to live a vibrant and active life and then die fairly quickly instead of experiencing the progressive and steady decline that many people go through, along with the various aches, pains, medications, and procedures that often accompany the last several decades.

Many anti-aging experts have used the term *health span* to define the years spent in good health, free from the disabilities and chronic illnesses that can come with aging. This is contrasted with *life span*, which is how many years you live, no matter what state you are in. What's the benefit of increasing life span if in your latter years you are incapacitated or otherwise unable to enjoy it? I think what most of us really want more than longevity is to increase our health span, and that, fortunately, is something we can most likely influence significantly.

Healthy aging is a natural process. It's a very gradual slowing down, as the body gets more miles on it and the brain gains wisdom. What isn't natural is premature aging, accelerated aging, aging caused by a junk-food diet, a sedentary lifestyle, industrial chemicals in the environment, or severe and chronic stress. Also largely preventable are the symptoms of premature aging: wrinkled and sagging skin, painful joints, foggy thinking and forgetfulness, fatigue, an inability to walk easily, climb stairs, or exercise, and the early onset of the so-called diseases of aging, from diabetes to heart disease, from autoimmunity to cognitive impairment.

There is no reason why we have to be dazed and confused for the last few decades of life. You are *not* destined to develop painful joints, diabetes, heart disease, or memory problems with age. The way you live today could stave off these signs of aging in the future. Only 20 to 25% of aging is genetically predetermined. The other 75 to 80% of how and whether you age is up to you.[6]

You can stay sharp, mobile, pain-free, and functionally healthy well into so-called old age. It's never too early to start staying young. No matter what shape you're in, you can reverse that process of internal degradation and support your body's natural processes of repair and detoxification. Your body is very good at rejuvenation—you just need to get out of the way. When you stop interfering with the powerful processes your body already has in place, like digestion, autophagy, collagen production, and immune function, you'll witness and *feel* dramatic changes. You'll slow or even reverse that accelerated aging process.

ASSESSING YOUR INTERNAL AND EXTERNAL AGE

How do you feel about this whole aging thing? What does it mean for you? What are your priorities? Your biggest concerns? What do you want to know more about, and what are you willing to do about it?

Think about this before you dive into the book. You may discover that you love parts of your life as an *older person* (even if that means you've just turned the ripe old age of 30), and you wouldn't want to go back to those less secure, more emotionally volatile younger years. But there may also be parts of aging you don't like, and that's totally natural, too. The wisdom and experience of age is a worthy trade-off, in my opinion, but still, it's hard not to long for the face, the body, the brain of a younger you. I get it. I feel that myself.

Think about what bothers you about aging and whether those things are actual signs of natural aging or could be signs of not taking care of your body as well as you might. Think about whether you are willing to change your lifestyle in ways that can promote autojuvenation.

Before we get into what you can do to reverse the signs of aging, let's assess where you are right now.

In each of these areas, check what applies to you and underline those things that especially bother you so you can get a clear picture of where you want to focus your efforts. The Younger for Life Program could prevent many of these symptoms of premature aging, and may be able to greatly reduce the discomfort and appearance of the items on these lists.

Skin

☐ Crow's feet

☐ Worry lines or *11* lines (those lines between your eyebrows when you furrow your brow)

☐ Nasolabial folds (those creases from the sides of your nose down to the corners of your mouth)

☐ Marionette lines (the creases from the corners of your mouth down your chin)

☐ Sagging face

☐ Drooping skin along the jaw line

☐ Skin hanging below the jaw

☐ Puckering skin on the neck

☐ Crepey or thinning skin on arms and legs

☐ Dropping bustline

☐ Loose upper-arm skin

☐ Drooping derriere

Energy

☐ Feeling tired in the morning, even after getting seven to nine hours of sleep

☐ Fatigue during the day

☐ Falling asleep in the afternoon

☐ Not having enough energy to exercise

Weight

☐ Weight gain without changing any habits

☐ Gaining weight around the torso (visceral fat)

☐ Weight-loss resistance: the things that used to work for weight loss don't work anymore

Posture

☐ Sloped shoulders

☐ A hump forming at the base of your neck

☐ Hunched-over posture

☐ Slumping when sitting because sitting up straight feels too difficult

☐ Walking stiffly

Muscles and Joints

☐ Shoulder pain

☐ Neck pain

☐ Upper- or middle-back pain

☐ Lower-back pain

☐ Pain shooting down one leg

☐ Painful hands

☐ Painful feet

☐ All-over joint pain

☐ Diagnosed osteoarthritis, bursitis, or bone spurs

Brain Function and Mood

☐ Brain fog

☐ Forgetfulness, such as not remembering where you put things or why you walked into a room

☐ Problems concentrating or paying attention

☐ Saying the wrong word

☐ Feeling down or blue more often, or depression

☐ Getting nervous more often, or anxiety

☐ Feeling "out of it"

☐ More frequent anger or irritation

Lab Values

☐ Elevated cholesterol

☐ Elevated or unstable blood sugar

☐ High blood pressure

☐ Inflammation as measured by an hs-CRP test

☐ Labs that suggest autoimmunity

Attitude

☐ Feeling old (no matter how old you actually are)

☐ Feeling like it's too late to do a lot of the things you wanted to do in life

☐ Feeling like you have no real purpose

☐ Isolating yourself, feeling like you don't have supportive relationships

☐ Loneliness

☐ A sense of giving up

☐ Losing interest in things that used to excite you

☐ Feeling sorry for yourself

☐ Not doing things to help other people (helping others is a surprising way to feel young again!)

So tell me: How are you feeling after checking some boxes? You may realize that your aging concerns are mostly cosmetic, or mostly health-related, or mostly mood- and attitude-related. Or maybe you checked a few boxes in all these columns. Now you have an idea about where you are. That's great news. Now you can keep track of your progress. It's time to take control of your destiny.

CHAPTER TWO:

WHY AND HOW WE AGE

Aging, as we've mentioned, is an inside and outside phenomenon. Very few people talk about this: most clinical doctors (dermatologists and plastic surgeons) focus on the exterior factors of aging, like wrinkles, age spots, and sagging skin caused by sun damage. This is why dermatologists are so focused on sunscreen and plastic surgeons on lifting what has fallen.

Now, it's true that sun damage just may be the most prominent outside factor that ages us. Just look at the skin of your butt and compare it to the skin of your neck and chest! I bet the skin of your rear is much smoother, tighter, and overall more youthful-looking than the skin of your chest, which may have some sun spots and fine lines. But aging is a *lot* more than sun damage.

Other things that increase the look of aging on our skin include:

- excessive movement of facial muscles, which can cause wrinkles to be deeper than they otherwise would be;
- using harsh products on the skin (like the old-fashioned

alcohol-based astringents), which can disrupt the skin's microbiome and result in damage;

- picking at the skin, which can cause scarring, making the skin look older;
- smoking and other unhealthy lifestyle choices. I can spot a smoker the moment they walk into my office. Their skin looks drier, duller, and more aged than it should.

But even that's far from the whole story. Focusing on the external causes of aging essentially treats the symptoms of aging, but not the cause. More important for *true* age are the inside factors. These include both genes and the effects of lifestyle habits such as diet, movement, sleep (or lack of it), and stress. Our lifestyle can profoundly impact how our genes express themselves, including whether certain genes are activated or repressed. The study of how this works is called epigenetics, and it's at the forefront of anti-aging and longevity medicine.

Prominent anti-aging scientists like Dr. David Sinclair, Dr. Valter Longo, and Dr. Nir Barzilai, among others, study how lifestyle impacts internal aging, with a focus on extending life and reducing or reversing disease. This is great, and significant, but even then, it's still only part of the story. It's true that if you only focus on the outside, you don't affect the cause of aging internally, but if you only focus on the inside, you still may not look as youthful as you'd like. Although many anti-aging scientists are brilliant, they aren't necessarily noted for looking 15 or 20 years younger than their actual age. The most powerful approach is to focus on both internal aging and external signs of aging, for maximum autojuvenating effect.

These two approaches are also connected. Even though you can impact the look of aging with external care, skin will look better if you are internally healthy. My friend Dr. Trevor Cates calls the skin *our magic mirror*. It tells us what is going on inside. Skin reflects health as well as age in many ways. Is it firm or sagging? Smooth or wrinkled? Flushed or sallow? Rough or soft? Broken-out or clear? These are all indicators of your internal health because many of the things that cause skin issues also cause internal issues.

Research supports this. A Danish study of 1,826 twins[7] from 2009 found that the younger-looking of two genetically identical twins usually lived longer, suggesting that a youthful appearance (external) correlates to more internal youthfulness. Of course, this doesn't take into account people who get cosmetic treatments.

The external signs of aging might be why you bought this book, but I hope to convince you that external appearance and internal health are intertwined, more so than we have ever realized before.

Here are five key lifestyle conditions that can cause accelerated aging, both inside and out:

1. Nutrient depletion
2. Inflammation
3. Collagen degradation
4. Free-radical damage (oxidation)
5. The buildup of cellular waste

Throughout the course of this book, we are going to tackle these five conditions. In doing so, you'll begin to see changes: smoother, tighter skin, better skin texture and thickness, a reduction in creping and wrinkling, and a healthier color, not to mention more energy, less pain, and that delightful sense of *feeling younger*.

SCIENCE WEIGHS IN

Let's take a step back for a moment and consider aging from a broader perspective, as we consider what we might want to do about it. Why do we age? Why do we die? And why do some creatures seem not to do either? I'm thinking of the so-called immortal jellyfish, which can revert back to an immature developmental stage in a stressful environment, essentially rebooting their aging process.[8] Or Brad Pitt, whose character aged in reverse in the movie *The Curious Case of Benjamin Button.*

We aren't jellyfish—or Benjamin Button, even if some of us think Brad Pitt is barely aging—but if we know what causes aging, we can gain clues about how to stop accelerating the process.

Let's start with something obvious we do multiple times every day: Food. Food is a particular target of anti-aging research. Or, I should say, the lack of food is. When scientists reduce the amount of food they give laboratory mice by 30 to 40%, the mice live longer. This has also been demonstrated in worms and insects and is being studied in primates. What we've learned is that caloric restriction slows down pretty much all normal age-related changes in the animals in which it's been studied.[9] A 2022 study[10] looked at this phenomenon even more specifically and discovered that:

- when mice were allowed to eat as much as they wanted, whenever they wanted, they lived an average life span;
- when mice were calorie-restricted by 30% but could eat *whenever* they wanted, they lived about 10% longer;
- when mice were calorie-restricted and were only allowed to eat during the inactive periods of their day, they lived about 20% longer;
- when mice were calorie-restricted but were only allowed to eat during the most active part of day and were not allowed to eat for at least 12 hours per day during the inactive part of the day (such as when they would naturally be sleeping), they lived 35% longer!

Of course, we can't say for sure that the same thing would apply to humans (and perhaps not incidentally, this study only looked at male mice), but it's interesting to contemplate whether something as simple as reducing calories and only eating during the day, while fasting for at least 12 hours overnight, might actually dramatically increase life span—and health span. The mice in this study showed actual genetic and metabolic changes associated with reduced inflammation, better blood-sugar stability and insulin sensitivity, and delayed progression of diseases, like getting cancer much later in their life cycles than mice who ate freely.

Just imagine what that could mean. Let's extrapolate from mice to humans (with the full understanding that this is just a hypothesis and not proven). The average human life span is about 73 years. If you added 35% to that, it would bump human life span to 98 years old! If those extra years were also characterized by low inflammation, stable blood sugar, and a healthy insulin response, and were free from the chronic diseases that so often afflict people when they get older, that would be pretty amazing. Slowing down aging could be as simple as eating less.

Another theory is that we have a set amount of energy and when we use it up, life is over. The ancient Hindus believed we get a certain number of breaths,[11] and when we use them up, we die. This may be why some yogis have for many centuries practiced extreme slowing of the breath. Could you slow aging by slowing your breath, such as by meditating daily?

Another theory is that life span is related to the number of times your heart beats. A Danish study that followed nearly 3,000 people for 16 years showed that, even correcting for fitness level, those with naturally higher heart rates tended to die sooner than those with naturally lower heart rates.[12] Specifically, study participants with a resting heart rate between 71 and 80 beats per minute were 51% more likely to die during the course of the 16-year study than those whose resting heart rates were lower than 50 beats per minute. Those with resting heart rates of 81 to 90 were twice as likely to die during the study, and those with resting heart rates above 90 were three times as likely.

Overall, most people get a little more than 2 billion heartbeats in a lifetime. Could you slow aging with regular exercise, which lowers resting heart rate? The idea that people with higher heart rates use up their heartbeats more quickly than those with slower heart rates, thereby causing them to die sooner, is called the Rate of Living Theory of Aging[13]—the idea that metabolic rate determines life span. This kind of makes sense. For one thing, the body's metabolism generates waste as a by-product of the energy it metabolizes, and this waste, in the form of reactive oxygen species or free radicals, can damage tissues. So wouldn't higher metabolisms generate more waste, damaging the body faster? Higher heartbeat, more breaths, faster metabolism, shorter life? Could we slow aging by lowering our metabolism?

Maybe. There is plenty of evidence in the animal kingdom to refute this. Many smaller animals with faster metabolisms don't live as long as some larger animals with slower metabolisms, but this isn't always the case. Take birds, for example. They have double the metabolic rate of mammals that are about the same size as they are, but they live about three times longer. So while the rate at which you burn energy may be related to aging, it's likely not a direct correlation.

But here's another twist—that Danish study about heart rate and longevity was only in middle-aged men. Does it apply to women? We don't know because that hasn't been studied. (Come on, researchers, your studies are sorely lacking!) However, there is another interesting hypothesis that is specifically about women and longevity: the grandmother effect.[14]

This hypothesis is about why human women tend to live so long after menopause—indeed, women tend to live longer than men.[15] The theory says that because grandmothers were (and continue to be) so instrumental to the survival and success of their grandchildren, this allowed those grandchildren to survive longer and also allowed their children to reproduce more. This led, over time, to more success in family lines with long-lived women, passing down the genetic traits of female longevity. (I must credit my father-in-law, who went askew of this theory when he spent thousands of hours helping take care of my children when Amy and I were working or away. Thank you, Jim! I'm sure you know of several grandfathers who do the same.) Maybe we can slow down aging, live longer, and influence future generations just by having a purpose and remaining useful.

To get more technical, another interesting theory of aging is the telomere theory. Telomeres are small portions of DNA located at the ends of chromosomes. They are often compared to the wrapped ends of a shoelace. Whenever a cell divides, the DNA unwraps, and the genetic information is copied. The telomere protects the DNA when the cell divides and is not itself copied, but a portion is clipped off with every cell division.

According to telomere theory, this gradual clipping off will continue until the telomere is gone. When that happens, the cell can no longer divide and begins to age. Cells that can no longer divide are called senescent cells, and although these cells can still function, it's believed that they may play a pivotal role in accelerating the aging process.

Research has revealed that older people do indeed have shorter telomeres.[16] We also know that shorter telomeres are associated with various diseases that usually occur in older people, like cardiovascular disease, hypertension, type 2 diabetes, and osteoporosis.[17] This makes sense in a way, but it's also currently unclear exactly how much telomere length directly impacts aging. Is it a process that causes aging, or is it a result of aging? And exactly how do senescent cells play a part in all the multiple facets of the aging process?

Until we know more, I think it's important to pay attention to the environmental factors that accelerate telomere degradation and are associated with shorter telomeres (smoking is one example) but also to realize that it is likely just one of many factors related to aging.

There are many other theories of aging with complicated names and even more complicated explanations, like the stem-cell theory of aging, the replicative senescence hypothesis, the neuroendocrine theory, the cross-linking theory, the immuno-logical theory, the somatic DNA damage theory, and the mito-chondrial theory. It's beyond the scope of this book to go into these, but it's good to know that scientists are on the case.

All these theories have their interesting points, and I suspect they all have some truth to them. But let's focus on what you can do in your life to directly impact the aging process. Science has shown that making the changes I recommend in upcoming chapters—to your diet, lifestyle, and exposures—can extend your health span and slow down and even reverse signs of aging.

Remember those calorie-restricted mice? The ones who lived the longest didn't eat as much and only during their active hours, and the result was lower inflammation, better blood sugar and insulin control, and a better immune response that kept them from developing chronic diseases until the very end of life. These are good goals for those who want to age better and stay alive longer. Do what lowers inflammation, keeps blood sugar and insulin stable, and strengthens the immune system.

There are still many unanswered questions about aging, but I've based this book around what we currently know and, es-pecially, what we can do about it. We can't change our genet-ics, but we can optimize the bodies and lives we have. So how do we do that?

BEGIN WITH THE SKIN

Let's begin with the skin, that magic mirror that indicates how we are aging on the inside. I like to use the skin as a point of entry for lifestyle changes because of the skin's reflection of both external and internal health. In dermatology, there are seven signs of skin aging:

1. Fine lines and wrinkles, like crow's feet and the *11* sign (those vertical lines between your eyebrows)

2. Dullness
3. Uneven tone
4. Dryness
5. Discoloration
6. Rough texture
7. Visible pores

There are four major causes of skin aging resulting in these signs, and they are the focus of the Younger for Life Program you'll learn about in the next chapter. In order to effectively improve our skin's appearance from the inside out, we need to address each of these four causes.

1. COLLAGEN DEGRADATION

Our skin is made up of about 75 to 80% collagen, and that collagen, which gives our skin structure, firmness, elasticity, and suppleness, breaks down over time, getting about 1% thinner every year. Collagen is a very complex (and fairly large) protein, so it's important to eat enough protein in order to help your body rebuild your collagen. Skimping on protein may hasten collagen degradation, especially with age.

But in addition to collagen getting thinner with age, it can also undergo degradation due to external factors, like sun damage and pollution, and internal factors, like what we eat. One particularly skin-relevant dietary factor of skin aging is sugar. When sugar molecules bond to collagen and elastin fibers in the skin, they deform them. This process, known as glycation, causes the skin to look older. Coincidentally, these sugar-protein hybrids are called advanced glycation end products, otherwise known as AGEs.

2. OXIDATION AND FREE-RADICAL DAMAGE

One of the primary causes of physical breakdown (and the basis of one of the major theories of aging) is oxidation and free-radical damage. I talked about this a few pages back, but let's look at it more closely. Free radicals are reactive oxygen species produced by the body as a by-product of metabolism but also by external means like UV exposure, smoking, junk food, and environmental exposures. These free radicals damage DNA, as well as collagen and elastin in the skin.

The body naturally produces antioxidants to neutralize free radicals, but if there are too many free radicals for the body to manage, the body enters a state called oxidative stress. The condition refers to when the body is being attacked by more free radicals than it can naturally handle. The result is DNA damage, along with premature aging. We can prevent or treat this by minimizing exposures that increase free radicals and maximizing antioxidants in the diet and on the skin. Our lifestyle choices, such as smoking or eating a lot of processed foods, can increase free-radical production.

3. CHRONIC INFLAMMATION

Another internal and external source of aging is chronic inflammation. Inflammation is a natural response to harmful stimuli. Acute inflammation can be a good thing. Getting a chemical peel or a laser treatment creates acute inflammation, which helps our bodies heal and fight off infection, or in this case, results in smoother and tighter skin. So acute inflammation can be a necessary and even beneficial process. It's when inflammation becomes chronic that problems occur.

Chronic inflammation can damage skin over time by causing skin conditions like acne, rosacea, and eczema. It can also stress and damage the endothelium, which lines all the veins and arteries of the body and can lead to heart disease, stroke, high blood pressure, metabolic diseases like diabetes and obesity, and just generally feeling old. The lifestyle choices we make can reduce inflammation or make it worse.

4. REDUCTION OF AUTOPHAGY

Autophagy is the natural process the body uses to clean up and remove dead and damaged cell components to make room for healthy cells to flourish. This process slows down with age (like so many other beneficial body processes), and the accumulation of damaged cells and cellular debris can accelerate aging. However, we have several effective strategies to support autophagy, especially intermittent fasting and calorie restriction.

I argue that these four processes—collagen degradation, free radicals, inflammation, and suppressed autophagy—are the main causes of aging skin as well as internal aging and are the exact targets of autojuvenation. Each one can be addressed through how you treat your own body—with the right kind of actions, free radicals can be neutralized, inflammation can be cooled, and autophagy can be ramped up. This is autojuvenation at work! Attacking aging from each of these four perspectives is at the heart of this book's protocol.

You're going to look younger, yes, and you're going to feel younger, too. This will be happening in many different ways at the same time. For instance, an anti-inflammatory diet may impact your telomere length, fight free radicals, and promote longevity genes, while eating nutrient-dense foods at certain times during the day can fight free-radical production while also helping to promote the process of autophagy.

Now that you understand some of the science behind aging, you will be better able to see why the Younger for Life Program includes what it does. I've got solutions for all of these issues, through lifestyle triggers for autojuvenation.

Meet the Younger for Life Program.

CHAPTER THREE:

THE YOUNGER FOR LIFE PROGRAM

One request that I receive over and over again is to create a truly holistic anti-aging program that is scientifically supported, holistically minded, and strongly effective inside and out. People often ask me, "If you're a holistic plastic surgeon, can you teach me what I can do holistically so I don't need to see you anymore?"

This is how I came up with the idea to use autojuvenation as the basis of a program with these patients and followers—and anyone else who would be interested—in mind.

I've striven to create an anti-aging program that is 1) simple and manageable, and 2) covers *all* the things we can do to trigger autojuvenation: diet, supplements, skin care, and lifestyle.

I must thank my social media followers, especially the ones on Instagram and YouTube, who, when polled, overwhelmingly asked me to write a book on a holistic anti-aging program over a guide on how to have plastic surgery using a holistic approach. You have my endless appreciation, and here it is!

I believe that if you're here in this life, you might as well live it! If you can have a body that feels younger than your years, a face that looks younger than your years, and a brain that is as wise as your years, you'll have the best of all possible worlds.

WHAT'S INVOLVED IN THE YOUNGER FOR LIFE PROGRAM

The Younger for Life Program is comprehensive because aging is complex and multifactorial. It's a way to tackle aging from many different directions at the same time. The first target is diet.

What you eat can slow aging through five different autojuvenative mechanisms:

1. Nourishing your body
2. Cooling inflammation
3. Firming through increasing collagen production
4. Healing from oxidative stress/free-radical damage
5. Cleansing the body of dysfunctional cells by increasing cellular turnover and supporting the cellular cleanup process called autophagy

To achieve these goals, the program promotes autojuvenation with a diet plan, a supplement plan, a skin care routine, and a jump start protocol to pull it all together. I'll also offer lifestyle tips and advice on procedures to supplement your plan.

THE YOUNGER FOR LIFE DIET PLAN

What you put in your mouth has a major impact on the appearance of your skin and on the health of your body, making both your outside and inside more impervious to the effects of time. The Younger for Life Diet focuses on five primary strategies that directly attack the aging process:

1. *Nourish:* Choosing foods that nourish the skin (and the whole body). Nutrient density is the foundation for feeding the cells to avoid premature aging and decline. The best way to achieve this is to eat mostly whole foods in as wide a variety as possible, to get the most nutrients— protein, carbohydrates, healthy fats, minerals, and vitamins, especially C and E, and the B vitamins.

2. *Cool:* Choosing foods that reduce inflammation, including foods that promote microbiome diversity and give an advantage to friendly flora (like fermented foods rich in probiotics, and fiber-full foods rich in prebiotics), and eliminating age-accelerating foods and those that create AGEs which can prematurely age skin.

   Think colorful fruits and vegetables, which reduce wrinkles on the outside and bolster immune function on the inside. Add lots of herbs and spices. Those colorful fruits and vegetables as well as herbs and spices are full of phytonutrients like carotenoids and polyphenols. We'll also focus on anti-inflammatory healthy fats, especially omega-3 fatty acids from seafood and seeds, and monounsaturated fatty acids from almonds, avocados, olives, olive oil, hazelnuts, and macadamia nuts.

3. *Firm:* Choosing foods that encourage healthy collagen production, which can help to combat the loss of collagen that can happen with age. Since skin is made up mostly of collagen, reinforcing and strengthening that collagen can have a dramatic effect in helping firm and plump the skin.

   In youth, our skin's collagen is made of tightly packed fibrils, kind of like the logs of a log cabin, but with aging, the collagen can become degraded. The logs get worn and may even fall apart. Many anti-aging treatments today focus on rejuvenating collagen (like collagen supplements, retinol creams, and laser treatments), which rebuild and rearrange the logs of that cabin, but a major way to shore up collagen is to eat sufficient amounts of protein.

4. *Heal:* Choosing foods that reduce oxidative stress by neutralizing free radicals. These foods protect against pollution, the sun, and other internal and external sources of wear and tear. As bodies age, these processes become less efficient, which can create reactive oxygen species, most commonly free radicals. Just as metal can rust, so too can our cells develop corrosion via a similar type of reaction.

   Our bodies' best defense is antioxidants, which neutralize free radicals by donating an electron to them. Our bodies create a certain amount of antioxidants, but increasing the number of antioxidants in your diet can give your body a much-needed infusion to help quell oxidative stress. It's like bringing in the cavalry during a battle.

5. *Cleanse:* Eating strategies that increase autophagy, or cellular turnover, including an efficient and rapid removal of damaged cells and cellular organelles to make room for new, fresh cells. It's cellular housecleaning. These strategies include calorie restriction and/or intermittent fasting (you'll learn how to find the right eating and fasting windows for your particular preference and lifestyle), as well as eating and restricting certain kinds of food to optimize autophagy. In the context of this plan, you can optimize autophagy while still enjoying yummy things like butter coffee, guacamole, olive oil, walnuts, and salmon.

This part of the program will kick off the process of autojuvenation. You'll feel more energy, less pain, and the anti-aging effects will show on your face and in the way you move. You'll also likely have a better mood.

Autojuvenated

L.D. is 45 years old and works as a project manager. Prior to starting the Younger for Life Program, L.D. was eating a standard American diet, having about seven alcoholic drinks per week, and suffering from chronic constipation, for which she was taking a daily dose of MiraLAX. She was not taking any supplements before starting the program and had a very simple skin routine of cleansing, moisturizing, and sunscreen. Her goals were to decrease age spots and fine lines and increase moisture in her skin. She didn't realize how good the program would make her *feel*.

After three weeks, L.D. told me that her skin looked fresher and felt softer, and that people mentioned she was looking at least two or three years younger. Her overall texture and wrinkles improved, and she felt better about herself and the choices she was making at mealtime. She also lost 3.2 pounds and said she really enjoyed the diet and found it much easier to follow than she expected. She especially loved the grass-fed beef tacos, which friends and family also enjoyed, as well as the oven-roasted vegetables with pistachio pesto. The skin care routine was also easy and really effective for her. Overall, she raved that this was a great program and helped her achieve a lot of discipline in her diet and skin care routine.

After two months, L.D. had re-introduced gluten and dairy into her diet, but she wasn't consuming as much as she had previously. She kept up the skin care routine and continued to see improvements in suppleness and texture. She reported that her friends noticed the improvement, saying she looked younger, glowing, and fresh, and couldn't believe she wasn't wearing makeup! L.D. continued to notice internal benefits as well, telling me that the diet made her feel better overall and improved her digestion.

THE SUPPLEMENT PLAN

The next part of the Younger for Life Program will show you how you can use supplements to augment and speed up the process of cellular rejuvenation by fully covering your nutritional bases, adding anti-inflammatory and antioxidant boosters, maximizing collagen production, and encouraging cellular turnover. Supplements can turn up the volume on all the good things you're doing with your diet, so you can help your body work and fill in any gaps you might not be getting from your food. My simple anti-aging supplement protocol won't have you gulping down handfuls of capsules (gag!). It's a streamlined approach specifically targeting premature aging and supporting autojuvenation.

THE TWO MINUTES, FIVE YEARS YOUNGER SKIN CARE ROUTINE

Next comes my quick and easy Two Minutes, Five Years Younger Skin Care Routine that can make your skin look and feel younger. This is primarily an external approach, but since skin is the barrier between you and the outside world, taking care of it is a great way to slow the aging process externally. I'll show you how you can nourish, cool, firm, heal, and cleanse your skin using the right skin care products, which is so incredibly important. I'll also talk about my concerns with many commercial skin care lines.

In the morning, you'll cleanse, use an antioxidant serum, and end with a sunscreen, to keep your skin clean, treated, and protected from the onslaught of sun and pollution. In the evening, you'll cleanse again, treat your skin with a retinoid, and use a brightening cream and/or moisturizer (both are optional). Two or three times a week, you'll exfoliate. And that's it! Simple, streamlined, and totally doable. You won't be standing in front of the mirror for hours every morning and evening. (Unless you want to, of course—feel free to admire yourself as you begin to look and feel younger!)

This section also includes some extra DIY treatments you can make and apply at home.

THE JUMP START

The next part of the Younger for Life Program is the Three-Week Autojuvenation Jump Start. This is your chance to take everything you've learned about how to use food, fasting, and skin care to reverse the aging process and begin implementing it in your life in a way that will bring quick and visible results. This autojuvenating jump start is *powerful*. It contains two weeks of routines and menus so you always know exactly what and when to eat. This can reset your diet as well as your attitude.

THE LIFESTYLE ADVICE

The last part of the Younger for Life Program is about the other things you can do in your life—physical, mental, and more that science says can actually reduce stress, improve mood, and change the body and the brain in ways that will make you feel younger, more energetic, more positive, and even happier. This is primarily an internal approach, but the changes you experience will be dramatic. I'll talk about yoga and meditation (which can change your whole life—I'm living proof), stress management, how to think and act and move more like a young person, and I'll even help you figure out what your purpose and mission are in life, because having something to live for is what it's really all about.

MINIMALLY INVASIVE PROCEDURES

Finally, for those who are interested, I'll talk about some of the minimally invasive, super safe, and affordable procedures you can do, if you really want to go beyond lifestyle adjustments. If I had my way, most people would avoid the big plastic surgeries in favor of living a healthy life, but if you truly want to get a little Botox, a bit of filler, or some of the newest technologies that can finesse your lifestyle efforts, I'll tell you everything you need to know. Heck, I do them, too.

And that's it! Are you with me? Great! Then, let's get started with the most basic and probably the most impactful thing you can possibly do every single day to look and feel younger and cure premature aging: eating the right food.

PART TWO:

REVIVE WITH THE
YOUNGER FOR LIFE PROGRAM

CHAPTER FOUR:

EAT TO NOURISH

Kathy is a 43-year-old mother of two who didn't take care of herself until, at the age of 40, she was diagnosed with ulcerative colitis. She decided to make major changes in her lifestyle so this new diagnosis wouldn't ruin her life. A ten-year cigarette smoker, Kathy quit smoking cold turkey and changed her diet completely. She gave up soda pop and energy drinks, replacing them with purified water. She reduced the amount of processed carbs and gluten she ate and completely cut out fast food. She began eating more colorful fruits and vegetables, became choosier about the meat she ate, and majorly curtailed her dairy intake.

Three years later, the adult acne she'd suffered with for years is gone. Her skin feels smooth and hydrated, and the dark circles under her eyes have faded. Her dental health is better, she has more energy than ever before, she sleeps soundly at night, and her mood is more pleasant and steadier than it had been since having children.

Kathy's story is a common one. I can't tell you how many times

I've heard that making simple dietary changes have changed people's skin, not to mention their energy and overall health, in profound and amazing ways. What we eat truly has a massive impact on the health, appearance, and aging of our skin. It's the obvious starting point for the Younger for Life Program.

EATING TO NOURISH

The purpose of food is nourishment. We all know this, don't we? Food can sometimes be entertaining, quell boredom, or be the center of a family gathering or social activities, but your body functions because you eat. The proteins, carbohydrates, fats, vitamins, and minerals, as well as the antioxidants and other anti-inflammatory and healing elements in food, especially in plants (we have only begun to identify all the healing compounds in plants), are what our bodies use to stay alive, remain healthy, and thrive.

What we eat provides the substrates we need to build muscle, regenerate skin, heal from injury and disease, and fight off infections from bacteria and viruses. Food gives us the nutrients to run our livers and kidneys, our brains and digestive systems, and all the complicated glands and processes involved in hormonal balance. With food, we can either squeak by or flourish. Without food, we can't live at all.

This is anti-aging level one. It gets more complicated, interesting, and targeted than this, and we will get there in this chapter. But first, you have to hit the basics. You may know what you're supposed to be eating, but are you doing it? Are you actually getting enough nutrition so that your body has what it needs to function? Many people aren't, and that is part of the reason why premature aging and diseases set in.

I think you already know that you need to eat high-quality protein, complex carbohydrates, and healthy fats each day. You need balanced meals with a lot of variety so your body can extract everything it needs. Don't start trying fancy, weird, or specialty diets until you establish a dietary foundation. But in case you aren't sure where to start, here are the rules for adults, to ensure they are getting nourished.

MACRONUTRIENTS

The three macronutrients are protein, carbohydrates, and fat. These are the nutrients so many fad diets play around with, claiming you should eat low-fat high-carb (common with people who avoid animal products or who are trying to avoid or repair heart disease), or low-carb high-fat ("keto" diets, often used for weight loss), or high protein (common with bodybuilders and other athletes), just to name a few of the variations on macronutrient themes. There might be a reason—medical, ethical, functional—to play around with macronutrients, but most regular people need a balance of these. About 50% of calories from carbohydrates, about 20% from protein, and about 30% from fat.

You might be wondering about exact amounts, such as how many grams of each you need. That depends on how much you eat. Let's say you eat 2000 calories per day, which is a general target for active healthy people. Considering that 1 gram of fat has 9 calories and 1 gram of either protein or carbs has 4 calories, you would need to eat about 250 grams of carbohydrates, 100 grams of protein, and 66 grams of fat each day, to hit those macro percentages.

You can calculate this using any of the many nutrition apps out there. You enter what you eat, and these apps can tell you how your macronutrients come out.

Of course, this doesn't consider the quality and source of your protein, carbohydrates, and fat. That's next-level.

MICRONUTRIENTS

Micronutrients are your vitamins and minerals, and the truth, according to the US Department of Agriculture, is that most adults in the States do not get enough of the vitamins A, B_6, C, D, and E, or the minerals calcium, potassium, magnesium, and iron required for good health. Fiber is another important food-based element most people don't get enough of—most adults need at least 28 to 30 grams per day.

Could you take a multivitamin and fiber supplements to cover those bases? You could, but if you really want to make sure you're actually absorbing those vitamins and minerals, the best way to get them is from food. Here's what you need for a healthy diet, or the RDA (recommended daily allowance), according to the USDA.[18]

VITAMINS

Vitamin A (and beta-carotene): about 700 mcg per day for women and 900 mcg per day for men.

Riboflavin (vitamin B_2): about 1.2 mg per day.

Niacin (vitamin B_3): about 14 mg per day for women and 16 mg per day for men.

Pantothenic acid (vitamin B_5): about 5 mg per day.

Pyridoxine (vitamin B_6): about 1.3 mg per day.

Folic acid or folate (vitamin B_9): about 400 mcg per day.

Cobalamin (vitamin B_{12}): about 2.4 mcg per day.

Vitamin C: about 100 mg per day.

Vitamin D: about 15 mcg or 600 IUs per day.

Vitamin E: about 15 mg per day.

Vitamin K: about 90 mcg per day for women and 120 mcg per day for men.

MINERALS

Calcium: about 1,000 mg per day and up to 1,200 mg per day for women over 50.

Chromium: about 20 to 30 mcg per day.

Iron: about 18 mg per day for women prior to menopause (after, about 8 mg per day) and about 8 mg per day for men.

Magnesium: about 320 mg per day for women and about 420 mg per day for men.

Potassium: about 2,600 mg per day for women and 3,400 mg per day for men.

Selenium: about 55 mcg per day.

Sodium: less than 2.3 grams per day (about a teaspoon of table salt).

Zinc: about 8 mg per day for women and 11 mg per day for men.

Not all nutrition tracker apps will get this detailed, but some will, so look around. You can see, as you track, where you are getting enough and where you are falling short, according to these recommended daily allowances.

WHAT A DAY OF EATING COULD LOOK LIKE

Rather than give you a strict eating plan, I want to emphasize that *variety* is the key to maximum nutrient diversity, and choosing foods with a lot of nutrients in them is the key to nutrient density, which simply means getting a lot of nutrition for the calories. Either change up your meals frequently (breakfasts could rotate between fruit and yogurt, eggs and veggies, oatmeal and nuts, or a green smoothie on rushed days, for example), or if you always have fruit and yogurt, vary the types of fruits you eat based on what is fresh and in season. If you always have a salad for lunch, change up the veggies and vary your protein (chicken, beef, black beans, tofu, salmon, shrimp). Variety not only provides a wider range of nutrients but keeps you from getting bored with your meals.

The Autojuvenation Jump Start will take you through an actual meal plan and recipes specifically designed to nourish and more, so if you want to get a head start, you can flip to the

Jump Start or peruse the recipes starting on page 196. They are designed to go with the Jump Start, but they are delicious, and you can make them—or use them for inspiration—at any time, even when you aren't doing the Jump Start.

CHAPTER FIVE:

EAT TO COOL INFLAMMATION

Once you've got nutrition in order, you can start to target the more specific issues related to premature aging, and first on the list is inflammation. Inflammation is a fairly general term for the body's response to a harmful stimulus, like an injury. When your sprained ankle swells or the scrape on your knee gets red around the edges, that's inflammation fulfilling its important function. This is called acute inflammation, and it's meant to be temporary. The swelling, heat, and redness sound an alarm, so your body knows to send healing resources to the injured area.

However, with constant low-level insults to your system (stress, pollution, processed food, alcohol, drugs, and did I mention *stress*?), you could develop chronic inflammation. This kind of inflammation is more harmful than beneficial because it never stops aggravating your skin, your digestive tract, your organs, and the lining of your veins and arteries. It's this kind of inflammation that likely leads to many common chronic diseases[19] because your body never gets the opportunity to fully heal. Few things will age a body faster than chronic inflammation.

But food can actually intervene and reduce chronic inflammation. There are four ways to target inflammation with the food you eat.

Autojuvenated

Donald is a 55-year-old male surgeon who grew up in a small town in Michigan. His parents were both doctors and often didn't have the time or energy to prepare healthy foods at home. Except for the Chinese place nearby, the handful of restaurants in town were all fast-food joints, so growing up, his diet consisted mainly of cheeseburgers, fries, corn dogs, and deep-fried sweet-and-sour chicken. He would wash it all down with soda pop.

Although he is a physician and had a small amount of nutritional education in medical school, Donald's adult diet consisted of many of the same things he grew up eating: fast food and sugary drinks. He was diagnosed with eczema and rosacea by a dermatologist, and his flaky, red face was speckled with the occasional pimple. Whenever I saw his face, the only words that ever came to mind were *chronic inflammation*.

When Donald reached his midfifties, he realized that some of his slightly older colleagues and acquaintances were struggling with health issues, and that if he didn't make changes in his diet, he would likely join them. The inflamed face he saw in the mirror each day was a sign of what was going on inside his body. But change is hard. Having eaten the same unhealthy diet for over five decades, he was unable to go cold turkey on the fast and deep-fried foods and soda pop.

What he was able to do, however, was to gradually reduce these foods, slowly transforming them from a daily staple to an occasional treat. This was all it took. His energy soared, and he felt better and healthier than he had 15 years earlier. Although he still had rosacea, eczema, and the occasional pimple, his skin looked less red and inflamed than it had in years. He is a great example of how even small changes can make a big difference.

EAT ANTI-INFLAMMATORY FATS

It might surprise you that one of the most powerful ways to fight inflammation, inside and on the skin, is by consuming more healthy, inflammation-lowering fats. Did you think fat was bad? Some types truly are and can cause more inflammation, especially trans fats, heated fats (e.g., the fats used to fry french fries and potato chips), and omega-6 fatty acids (when they aren't balanced by sufficient omega-3 fatty acids). These fats can cause accelerated aging and even lead to chronic-disease issues.

However, other kinds of fat have strong anti-inflammatory action. These come in two categories: omega-3 fatty acids and monounsaturated fatty acids. Both fight inflammation inside your body and on your skin, while also helping your skin to retain moisture better.

- **Omega-3 fatty acids**, like the kind in fatty fish such as salmon, mackerel, and sardines, as well as in seaweed and some kinds of seeds, intervene directly in the inflammatory process and are part of the body's natural defense against inflammation after an injury is healed. Because people often don't get enough omega-3 fatty acids, they may have trouble resolving inflammation after an injury. Help your body do what it's supposed to do naturally and give it sufficient omega-3 fatty acids!

- **Monounsaturated fatty acids** are prevalent in some plant-based foods, such as olives and olive oil, avocados, nuts, and seeds, and can help to quell chronic inflammation from inflammatory fats as well as the kind of inflammation common in people with an excess of body fat.[20] It can also help to raise your HDL cholesterol (the so-called good kind that can help to prevent heart disease). Note that meat and dairy products also contain some monounsaturated fatty acids, but research shows these may not have the same benefits (namely lower mortality) as plant sources.[21]

EAT FOODS TO SUPPORT YOUR GUT MICROBES

You have an unlikely ally in your fight against inflammation: bacteria. The bacteria in your digestive tract is especially good at reducing inflammation indirectly. As they digest the fiber and resistant starch that your own digestive tract can't digest, the little bugs in your gut produce metabolites that have anti-inflammatory effects on the macrophages that produce inflammation in your body. These metabolites include short-chain fatty acids and lipopolysaccharides.[22] They can have a noticeable effect not just on your immunity but on the way your skin looks.

My good friend, the dermatologist Dr. Whitney Bowe, author of the book *The Beauty of Dirty Skin*, is a major thought leader regarding the connection between gut health and skin health. She wrote one of the first articles ever published in the scientific literature on this subject,[23] which describes the gut-skin axis and looks at how the microbiome can influence acne. While you may think of acne as a problem of youth, many people continue to suffer with it throughout life or have a recurrence when hormones start to change again around the menopausal transition.

We've also learned that there is a strong association between small intestine bacterial overgrowth (SIBO), a condition in which pathogenic strains of gut bacteria overgrow up into the small intestine where they aren't supposed to be, and the incidence of rosacea, which is an inflammatory skin condition.[24] There are also links between inflammatory bowel disease (IBD), a gastrointestinal autoimmune condition, and inflammatory skin disorders like psoriasis, rosacea, and ectopic dermatitis.[25] What's more, researchers have found that oral probiotics can improve some cases of acne[26] and can even reduce UV-induced skin damage in mice.[27]

Your gut microbiome has a lot of influence over your health, digestion, and immune system. Some foods increase your microbiome's ability to lower inflammation, while others help inflammation-causing bacteria and yeasts to proliferate. Some of the best foods for supporting a healthy microbiome are fermented foods such as yogurt, kefir, kimchi, kombucha, sauerkraut, tempeh, and miso, which contain probiotics like those that live in your microbiome.

Yet, despite all this good evidence, and all we have learned about the valuable health benefits of a balanced gut microbiome with plenty of beneficial and anti-inflammatory gut bacteria, the standard American diet has all but eliminated fermented foods. Our meals have become essentially sterile, lacking both prebiotics—the fiber and resistant starch our good bacteria like to eat—and probiotics, the beneficial bacteria themselves. You might not be able to find fermented foods rich in probiotics at the local drive-through, but you can find them in stores if you hunt a bit. Once you get used to that sour taste, you may find you really enjoy these foods.

One of my favorite fermented foods is kimchi. This is a spicy pickled cabbage that is a staple in the Korean diet. It's the most popular *banchan* (traditional Korean side dishes served with every meal), and one that I love, but my wife and kids…not so much. If it's too spicy for you, try dipping it in water to wash off some of the spice without washing away the flavor and probiotics. This ancient Korean secret was something I enjoyed as a kid.

Kombucha is a probiotic-rich drink made from fermented black tea with sugar, but most of the sugar gets digested by the bacteria so it's overall pretty low in sugar and calories. It's a great alternative to soda pop, since it's sweetish and carbonated. It can contain a very small amount of alcohol, so I don't recommend it for children, but it probably won't be enough alcohol for you to notice any effects. I don't recommend having more than one cup a day, especially if it has added sugar. (I also don't recommend kombucha for anyone who is sensitive to caffeine, acidity, or alcohol, or anyone who is pregnant or immune-compromised.)

Kefir is another fermented drink, kind of like a drinkable yogurt, but with more probiotics than a typical serving of yogurt.[28] It can also be made from coconut water or coconut milk, so it can be dairy-free. Kefir is probably the most intensely probiotic-rich of all the probiotic beverages.

If you love sauerkraut on your brats, then you already love at least one popular fermented food rich in probiotics. Sauerkraut (like kimchi) is made from fermented cabbage, but (unlike kimchi) it's not spicy. Look for lacto-fermented brands that are made with actual bacterial cultures, not just soaked in vinegar.

Two fermented soy products that contain probiotics are miso (as in miso soup) and tempeh. Ideally look for organic varieties, since soybeans tend to be heavily sprayed and genetically modified (they are among the most prevalent GMOs,[29] or genetically modified organisms).

And what about yogurt, that popular food people most often think of when they think of probiotics? That depends. Many people are sensitive to dairy and have skin reactions to it, not even knowing this is the reason. That's why I typically avoid it and recommend others do, too. However, if you know it doesn't give you a problem and you like yogurt, I recommend organic yogurt made from the milk of grass-fed cows because it has higher levels of anti-inflammatory omega-3 fatty acids.

Greek yogurt is also rich in protein, which is good for collagen production, but it also contains a lot of casein, which is a common allergen. Some people find goat milk yogurt to be easier to digest than yogurt from cow's milk. You could also try plant-based yogurts. Many brands are now made from almond or coconut milk. Whatever yogurt you choose, plain and unsweetened types are best. Mixing in fresh fruit is the best way to sweeten your yogurt.

EAT FOODS TO FEED
YOUR GOOD GUT MICROBES

Another way to bolster your beneficial bacteria is by eating foods rich in fiber and resistant starch. These are the prebiotics I mentioned earlier that you can't digest, but that your beneficial bacteria *can* digest. Foods high in fiber and resistant starch will help these bacteria grow, so they can fight off and crowd out inflammation-producing gut bugs. The American Heart Association recommends eating 25 to 30 grams of fiber every day, but most people don't get nearly that much.

When you don't feed your beneficial bacteria, the inflammatory bacteria that thrive on sugar and saturated fat can gain a foothold and upset your microbial balance, so go for lots of vegetables and fruits, as well as whole grains (without gluten if you are sensitive—see my discussion about that on page 103) and seeds, for more fiber. Good sources of fiber include apples, beans, avocados, berries, and cruciferous vegetables like broccoli and cabbage.

Resistant starch is another kind of carbohydrate that your gut bacteria love to digest. One of the best ways to get more resistant starch, weirdly, is to cook high-starch foods like pasta, potatoes, rice, and corn tortillas, then let them cool in the refrigerator.

This causes them to form more resistant (indigestible) starch, so that when you reheat and eat them, you get more of the good stuff. Other great sources are oats, green bananas and plantains, corn and grits, black beans, chickpeas, and cashews.

REDUCE INFLAMMATORY FOODS

Inflammatory foods are typically high in refined sugar, refined grains, saturated fat, and heated fats (as in fried food). I'll talk more about these in the chapter on age-accelerating foods (page 97).

LOSE WEIGHT IF YOU HAVE A LOT TO LOSE

A little excess weight usually doesn't cause inflammation, but a lot of excess body fat can be inflammatory, and weight loss—even a small amount—can often bring down chronic inflammation significantly.[30] (This isn't a weight-loss book, but many of the principles in this book, if you implement them, could result in the loss of unwanted body fat.)

Water Does a Body Good

When it comes to beverages that support your body's natural rejuvenation processes, water is always my first choice. Water hydrates your entire body so everything works better. It plumps and smooths your skin and also helps you digest your food better. Are you getting enough to support your health? Ideally, I recommend drinking eight cups, or about 64 ounces, of water every day. If you are a small person, you may not need that much, and if you are a large person, an athlete, or have been out in the sun all day, you may need more.

Hydration matters for health. Dehydration can interfere with healthy metabolism and might even contribute to the development of degenerative diseases and a shortened life span.[31] A 2021 study of agricultural workers showed that dehydration was associated with acute kidney injury.[32] So drink up, people!

But not all water is created equally. I recommend drinking filtered water without the chlorine and fluoride in most tap water, as there is some evidence that these and other contaminants common in tap water may be hazardous to your health if ingested in large quantities.[33] In my house that is on a lake, we use the Aqua Tru triple reverse osmosis water filter. Our water comes out of a well, so it flows from the taps with a slightly brownish/yellowish tinge. I tell you, once we filter that water through our Aqua Tru, it's completely crystal clear and so, so refreshing! There are several kinds of triple reverse osmosis water filters available, so I encourage you to look for one for your home. It's a good investment.

AUTOJUVENATING FOODS:
ANTI-INFLAMMATORY FOOD LIST

For the best autojuvenating, anti-inflammatory action, eat these foods often. Please visit autojuvenation.com to download a free printable shopping list of these and other autojuvenating foods, which you can bring to the grocery store.

FOODS HIGH IN OMEGA-3 FATTY ACIDS

Chia seeds	Herring
Cod	Mackerel
Flaxseed (ground)	Oysters
Grass-fed beef	Salmon (wild-caught)
Halibut	Sardines

Seaweed
Tofu (choose organic if
   possible)

Trout
Tuna (ideally fresh, not
   canned)

FOODS HIGH IN MONOUNSATURATED FATTY ACIDS

Almond butter
Almonds
Avocados
Hazelnuts
Macadamia nuts
Olive oil

Olives
Peanut butter
Pecans
Pumpkin seeds
Sesame seeds
Walnuts

FERMENTED FOODS FOR BETTER MICROBIOME ANTI-INFLAMMATORY ACTION

Kefir
Kimchi
Kombucha
Lacto-fermented pickles

Miso
Sauerkraut
Tempeh
Yogurt

HIGH FIBER AND RESISTANT STARCH-RICH FOODS

Almonds
Apples
Artichoke hearts
Avocados
Barley (if you are not
   gluten-sensitive)
Black beans
Blackberries
Blueberries
Broccoli
Brown rice
Brussels sprouts
Cabbage

Cashews
Cauliflower
Chia seeds
Chickpeas
Corn (and corn tortillas)
Edamame
Flaxseed
Green bananas
Grits
Jicama
Lentils
Oatmeal
Pears

Peas
Plantains
Popcorn
Potatoes, cooked and
   cooled
Quinoa
Raspberries

Split peas
Strawberries
Sweet potatoes
Whole wheat pasta and
   bread (if you are not
   gluten-sensitive)

CHAPTER SIX:

EAT TO FIRM UP COLLAGEN

Collagen makes up about 75 to 80% of skin, and it gets thinner with age. But it's not just skin that's made of collagen. Of all the types of protein in the human body, collagen is the most prevalent. It has a matrix-like structure that holds us together, via our connective tissue, and it's the primary protein making up our bones, muscles, tendons, ligaments, and cartilage.

Collagen is made of amino acids—the building blocks of all proteins—and because it is so important for structural support,[34] losing it means losing structure and firmness. Less collagen means sagging of skin and weaker tissues, like tendons, ligaments, and cartilage that tear more easily, and weaker bones that are more likely to break. You can see why collagen is so important for fighting aging.

Foods that promote collagen production help give the skin and connective tissue its structure, strength, elasticity, and youthfulness. Choosing foods that encourage healthy collagen production can firm and plump skin, but they can also make it easier to maintain muscle and decrease the risk of injury with exercise.

Attacking collagen degradation is a focus of many anti-aging treatments available today, ranging from collagen supplements to retinol creams to lasers. These interventions rearrange the collagen fibrils, often with good effects, but the best way to rebuild collagen is to eat enough—and a wide variety of—protein.

So how do we get more and better protein? When people think of protein, they usually think of meat, and it's true that animal products are a good source of bioavailable protein. However, they are not the only source. I believe people eat way too much meat, and there is some compelling evidence that too much red meat is inflammatory.[35] It's worth noting, however, that many of the studies looking at this did not distinguish between conventional red meat, grass-fed red meat, and processed meat, and one study that did distinguish showed increased inflammation with consumption of mixed and processed red meat, but not with consumption of unprocessed red meat alone.[36] It may well be that processed meat is much worse than conventional red meat, and that grass-fed beef isn't actually inflammatory at all. My friends who eat a paleo-style diet (who strive to eat a natural human diet before processed food, and even before agriculture, when grain consumption increased) say that fresh grass-fed beef is one of the healthiest and most nutrient-dense foods, and they may be right.

In my view, people in the US still tend to overdo meat consumption. I personally prefer to eat vegetarian most of the time. I admire those who eat a totally plant-based diet, but I also think that diets are highly individual and different people thrive on different kinds of diets, so this is an individual decision. When I eat meat, I try to be choosy about the quality. I advise the same for you, and I also advise keeping serving sizes small—about the size of your palm, or a little less.

Does meat quality influence how well the protein in the meat promotes collagen growth and prevents collagen degradation? I've never seen any evidence of this. However, animals fed on cheap grains like corn (and even lower-quality feed), which are not their normal diets, tend to have more potentially inflammatory omega-6 fatty acids, and grass-fed, pastured, or wild-caught animals tend to have more omega-3 fatty acids in their tissues, so from an inflammatory point of view, organic, grass-fed, pastured, and wild-caught is probably best. If you can afford it, I recommend choosing higher-quality meats when possible.

Bone broth is a powerful way to accomplish two things: add collagen to your diet and soothe your microbiome. Bone broth is made by simmering animal bones and cartilage and adding vinegar and spices if you like. It's probably best to use organic bones for the purest, most nutrient-rich bone broth. The bones are simmered for 24 to 72 hours, sometimes in a slow cooker, or you can do it more quickly using a pressure cooker. The collagen from the bones and tissues melts into the broth and is transformed into gelatin.

You can tell if your bone broth is truly rich in collagen if, when you refrigerate it, it turns into gelatin. Simply heating it melts the gelatin and turns it back into soup, for a comforting and pleasant meal or snack that really can help with your collagen production. Although there are no published scientific studies that I know of to support this, many holistic health practitioners, including my good friend Dr. Kellyann Petrucci, have found that drinking bone broth can really improve the appearance of the skin, and there have been many studies that prove ingesting collagen can have beneficial effects on the appearance of the skin.[37]

Organic Meat and Fish

Are organic, grass-fed, pastured, and wild-caught meat and fish better? In many ways, probably. In some ways, definitely. As an animal lover, I choose grass-fed, pastured, wild-caught, organic meat and fish whenever I can, not only because of its superior fat content but because the animals are treated more humanely (or at least they are supposed to be). Happy animals may not provide more nutritious food, but it still puts my mind at ease.

There is also evidence that animals raised in a more natural environment are beneficial to *our* environment and eating less meat overall may help as well. There is some research showing that livestock creates significant greenhouse-gas emissions,[38] and conventionally raised animals are often injected with a large amount of antibiotics to ward off infection from unsanitary living conditions.[39] Those antibiotics in the meat go into us, so eating less meat, and better meat, is probably beneficial, not just for your health but for the environment and for the welfare of other living beings.

All that being said, there are other specific health benefits to choosing higher quality meat.

1. **Grass-fed beef** has more omega-3 fatty acids than conventional grain-fed beef, which is higher in potentially inflammatory omega-6 fatty acids.[40] Grass-fed beef may also have more conjugated linoleic acid (CLA) than grain-fed beef.[41] CLA is a fatty acid that's been shown to aid in weight loss in humans, specifically reducing the amount of body fat,[42] so grass-fed beef might even help you lose weight, if that is something you are trying to do. Finally, grass-fed beef has more antioxidant carotenoids than grain-fed beef.[43]

2. **Pasture-raised or organic chicken** is less likely to harbor antibiotic resistance than conventionally raised chicken.[44] In fact, *Consumer Reports* found that 97% of raw chicken breasts were contaminated with potentially deadly antibiotic-resistant bacteria, including salmonella, staph aureus, and enterococcus.[45]

Although there isn't any good evidence I've found that pasture-raised or organic chicken is more nutritious than conventionally raised chicken, there is no doubt pastured chickens have a higher quality of life. Even so-called free-range chickens may be crammed into a barn and never go outside, as the term *free range* legally only means the chickens have access to the outdoors, not that they are actually out there. *Cage-free* just means the chickens aren't in cages, but they can still be crammed into massively unsanitary indoor buildings. Don't be fooled by these two designations, as they may not mean much, at least not from the point of view of the chicken.

3. **Pasture-raised pork** is also less likely to harbor antibiotic-resistant bacteria,[46] and may contain more beneficial fats[47] and vitamin E,[48] than conventionally raised pork. In the wild, pigs typically like to eat fruit, vegetables, meat, grass, and brush. In the feedlot, they are mostly fed grain, corn, and soybeans.

4. **Wild-caught fish** is more likely to contain lower levels of potentially harmful chemicals like PCBs and dioxin[49] and possibly contain higher levels of anti-inflammatory omega-3 fatty acids.[50] Farmed fish has been widely shown to contain far more contaminants. The best option is wild-caught, cold-water fish like salmon and mackerel, which also have the most omega-3 fatty acids. If you won't eat fish and don't want to take fish-oil supplements, you can get omega-3 fatty acids from algae-based supplements.

You can also support collagen production with plant foods that are rich in protein. Nuts, like almonds, walnuts, pistachios, chestnuts, Brazil nuts, and hazelnuts, are packed with protein, healthy fats, and fiber, so they tackle anti-aging from multiple angles. Seeds often contain less fat than nuts but are packed with skin-healthy ingredients, like protein, some healthy fats, zinc, and a lot of nutrients. The best are sunflower seeds, chia seeds, pumpkin seeds, and hemp seeds.

Legumes, like lentils, chickpeas, kidney beans, cannellini beans, and navy beans, are great sources of protein, fiber, vitamins, and minerals. These foods are especially important if you don't eat meat. Soybeans are also legumes, and so are their products, like tofu and tempeh. However, as I've previously mentioned, a large proportion of soybeans are genetically modified. The jury is still out on the long-term health effects of genetically modified foods, but there does seem to be some cause for concern.[51] For this reason, I believe it's important to look for organic foods, if you have access to them and are willing and able to pay the higher price. If not, then don't stress about it. The actual food you eat is more important than whether it's organic or not. I'd rather you eat a conventionally grown apple than an organic cookie any day.

One controversy about legumes is that they contain antinutrients. These are compounds that can interfere with the absorption of important micronutrients like calcium, iron, and magnesium. The two main antinutrients associated with legumes are phytic acid and lectins, which are believed by some to cause damage to the digestive tract and block mineral absorption. However, these claims have largely been debunked, as long as legumes are properly prepared.[52] If you soak dried legumes for 12 to 24 hours and allow them to sprout before cooking, then boil them, these antinutrients are deactivated. If you add a little baking soda to the water, that can help neutralize the antinutrients even more. Soaking also allows the good bacteria to digest away many of the antinutrients through fermentation. Pressure cooking may also neutralize antinutrients. Canned legumes have typically already been prepared in a way that neutralizes the antinutrients.

In my opinion, as long as you properly prepare your legumes, they are a great component of an autojuvenating diet. People have been eating legumes for centuries. They are a natural food with many health benefits. I don't think you need to pay too much attention to people who claim that legumes are dangerous or try to scare you away from eating them.

EAT YOUR BETA-CAROTENE

Eating foods rich in beta-carotene, which is a precursor to vitamin A, is anti-aging in two ways at once: a double threat to wrinkles! For one, beta-carotene is an antioxidant; I'll go more into what these do in the next chapter. It's also particularly good for your eyesight, which is exactly why your mother told you to eat your carrots. Also, your body uses beta-carotene to make collagen. If it's a fruit or a vegetable and it's orange, you can bet it's full of beta-carotene.

Fruits and Veggies

It's not just me who thinks eating more fruits and vegetables will help you to age more slowly—although I do have my personal reasons to think so. My parents grew up in South Korea, eating a diet consisting of rice, fruits, vegetables, and fish, which they have continued to this day, albeit in Orange County, not Seoul. My dad is in his eighties and my mom is almost there, and both of them are poised to live well into their nineties. Just seeing how they and my grandparents ate, and how long they've lived, is evidence enough for me to believe that focusing on fruits and veggies is the key to longevity and a youthful appearance.

A lot of research backs this up. Studies have shown that people who eat more fruits and vegetables age more slowly than those who don't. One international study found a statistically significant slower aging process in people who consumed lots of vegetables, legumes, and olive oil, and more rapid aging in people who took in more butter, margarine, milk, and sugar.[53] Another study of over 4,000 women in the US between the ages of 40 and 74 showed that those with a higher intake of vitamin C (which is found in many colorful fruits and vegetables) was correlated with fewer wrinkles, while high intakes of fat and carbohydrates were associated with more wrinkles.[54] Remember, our skin is our magic mirror, and antioxidants may be an important part of that magic. So pass that fruit plate to autojuvenate!

AUTOJUVENATING FOODS: COLLAGEN-BOOSTING FOOD LIST

To keep your skin as firm, elastic, and young-looking as possible, emphasize these collagen–boosting foods in your daily diet. Please visit autojuvenation.com to download a free printable shopping list of these and other autojuvenating foods, that you can bring to the grocery store.

FOODS HIGH IN PROTEIN

Almonds, almond butter
Beef
Buffalo/bison
Chia seeds
Chicken breast
Cod
Eggs
Elk
Flaxseed
Ham

Hamburger
Lamb
Lentils
Peanuts, peanut butter
Pistachios
Pork chops or tenderloin
Protein powder
(especially those made with pea and/or rice protein)

Pumpkin seeds
Salmon
Shrimp
Soybeans, ideally
  organic
Sunflower seeds

Tempeh
Tilapia
Tofu, firm or extra firm
Tuna
Turkey
Yogurt

FOODS HIGH IN BETA-CAROTENE

Apricots (especially
  dried)
Cantaloupe
Carrots
Dark green, leafy
  vegetables
Mangoes

Nectarines
Papaya
Peaches
Pumpkin
Sweet potatoes
Winter squashes
  (butternut, acorn)

CHAPTER SEVEN:

EAT TO HEAL WITH ANTIOXIDANTS

People often confuse anti-inflammatory foods with antioxidant-rich foods, and that is understandable because there is a lot of crossover. Many foods contain antioxidants and also have anti-inflammatory effects. What's more, free radicals can cause inflammation, and inflammation may increase free-radical production, so the mechanisms are intertwined.[55] However, the two have different functions. To understand what an antioxidant is, first you need to know what a free radical is.

Your body is always doing work and expending energy through your metabolism and processes like digestion. Like anything that expends energy, it creates waste. These by-products of the body's energy expenditure are called reactive oxygen species, or ROS, which is a type of free radical that contains oxygen. They can also be generated by exposure to toxins like cigarette smoke, pollution, radiation, sun exposure, and certain medications, as well as by a processed-food diet, alcohol, fried food, inflammation, infections, a sedentary lifestyle, excessive mental stress, simple aging, or exercising *too much* (when you don't give yourself rest days or enough recovery time).[56]

These oxygen-containing free radicals are missing at least one electron, so they try to steal electrons from healthy cells, damaging the healthy cells' structure and/or DNA. This process is called oxidation. It's a natural process because free-radical production is a necessary by-product of normal bodily functions. Our bodies are built to handle free radicals. But when your lifestyle generates more free radicals than your body can handle (such as when you eat a poor diet, smoke, are exposed to pollution, or are under a lot of stress), free radicals can overwhelm the system, creating a state called oxidative stress. This is when free radicals can begin to cause visible damage, like a breakdown in collagen and elastin, which can make skin looser, thinner, and more wrinkled, not to mention internal damage of organs and blood vessels that can eventually result in chronic diseases.

Fortunately, we have an easy remedy for free radicals: antioxidants. Antioxidants neutralize free radicals by donating an electron without damage to your cells. To help your body manage the free-radical onslaught, eat a lot of foods rich in antioxidants. Brightly and deeply colored fruits and vegetables are your ally in this quest. Antioxidants are the very substances that give them their vivid colors. Some of the key substances in plant foods that are your best allies against free radicals are:

- **Vitamin C.** This powerhouse antioxidant is one of the most important. It's not just associated with fewer wrinkles but with better production of collagen and elastin in the skin. It is water-soluble, which means any extra will be harmlessly peed out, so you really can't get too many vitamin C–containing foods (although too much vitamin C in supplement form has been known to create digestive distress). That's the good thing about water-soluble vitamins. The bad news is you don't store it in your body, so I recommend eating at least two servings of vitamin C–rich foods[57] every day. Also note that vitamin C degrades in fruits and vegetables after harvest, so the fresher your produce, the more vitamin C it will contain. The best place to get fruits and veggies is the local farmers' market, where they are more likely to be quite recently harvested and therefore extra fresh and chock-full of vitamin C. Some skin serums also contain vitamin C, which is useful from the outside in, as well as from the inside out.

- **Vitamin E.** Vitamin E is excellent at protecting skin from UV radiation as well as preventing free-radical damage. Unlike vitamin C, which is water-soluble, vitamin E is fat-soluble, so you can store it in your fat cells. For this reason, it isn't as important to eat it every day, but I do recommend eating vitamin E–rich foods such as sunflower seeds, almonds, peanut butter, avocado, and dark leafy greens, at least a few times per week. You may also have noticed that you can find vitamin E in many skin serums.
- **Carotenoids.** These are plant pigments that give fruits and vegetables a red, orange, yellow, or sometimes a green color, and they also happen to be powerful antioxidants as well as anti-inflammatories (two for one!). The subcategories of carotenoids include lycopene, lutein, and zeaxanthin.
- **Lycopene.** This makes vegetables and fruits red (think watermelon and tomatoes), and it's one of the most potent of the antioxidants. Cooked tomatoes are the best source because the lycopene gets concentrated, and it becomes even more bioavailable to your body with a little oil because it's fat-soluble. So the next time you make pasta sauce, add a little olive oil. It's great for your skin, and some research suggests tomatoes might even protect your skin from sunburn[58] (although use sunblock anyway, I beg of you!).
- **Lutein and zeaxanthin.** These don't just slay free radi-

cals. They have also both been shown to protect the skin from UV exposure, especially to the eyes.[59] These carotenoids are most prevalent in cooked leafy-green vegetables, but they are also contained in salmon and eggs.

YOUNGER FOR LIFE SOURCES OF ANTIOXIDANTS

In addition to antioxidant-rich veggies, fruits, legumes, and nuts, some of the most potent sources of antioxidants come in herbs, spices, and beverages. Here are some more potent antioxidant foods.

- **Herbs and spices.** Surprisingly, herbs and spices (especially dried because they are concentrated) contain more antioxidants than most other foods—something that's often overlooked in a food product that most consider not to have any real nutritional value.[60] Whether it's basil or thyme or cinnamon or cumin, adding more herbs and spices to your food will do so much more than improve their flavor. They offer a big antioxidant punch, so sprinkle away! In case you were wondering, cloves have the most antioxidants of all the spices.
- **Dark chocolate.** When I say *dark*, what I mean is chocolate that is at least 70 to 80% cacao. The more cacao, the higher the antioxidant content. One study showed that the antioxidant activity from dark chocolate was greater than that of red wine or green tea![61] Good-quality chocolate will show the percentage of cacao on the package. Most low-cacao chocolate won't. The more you get used to pure dark chocolate with its less sweet taste, the more it will grow on you. As a kid and even in early adulthood, I loved milk chocolate. As I've gotten older (and more knowledgeable about health), I've converted to dark chocolate, and now milk chocolate tastes waxy to me. You'll know you are a

true chocolate lover when you can enjoy raw cacao nibs, which are pure pieces of the cocoa bean with no sugar added. Try adding them to smoothies for extra antioxidant power *and* a pure, heavenly chocolate taste.

- **Green tea.** Green tea is filled with two powerful antioxidants: catechins and polyphenols.[62] It's probably the most powerful antioxidant drink—the catechins in green tea specifically have been shown to significantly enhance the effects of vitamins C and E, increasing antioxidant activity more than the vitamins or the catechins alone.[63]

  The next time you're looking for the perfect beverage to go with your (vitamin C–rich) fruit salad or your (vitamin E–rich) peanut butter sandwich, green tea may be the ticket. It's also a great antioxidant on its own.

- **Matcha.** Matcha is like super-powered green tea because it consists of actual green tea leaves dried and powdered, then mixed with hot water. When you drink matcha, you are actually consuming the leaves, not just the water the leaves were steeped in. Matcha has 2 to 10 times more antioxidants than traditional green tea, but it also has more caffeine, so it can be both stimulating and dehydrating. It also has a very strong taste that not everybody loves. There are also some concerns that matcha produced in China may contain lead, so look for matcha made in either the United States or Japan. I recommend no more than one cup per day for those who enjoy it.

- **Coffee.** Coffee is full of antioxidants (coffee drinkers, rejoice!), including melanoidins and chlorogenic, ferulic, and caffeic acids.[64] People who drink coffee regularly seem to have a reduced risk of heart disease, dementia, certain cancers, and type 2 diabetes. However, it can raise blood pressure in some people, and it can stain your teeth if you don't brush your teeth after drinking it.

  Honestly, if you tolerate caffeine, the worst thing about

coffee is all the junk people put in it. (I'll pass on the double-mocha caramel latte with whipped cream and syrup, although my daughter unfortunately loves these desserts-in-a-cup!) Creamer, sugar, flavors, and milk can be problematic, especially if you load up regularly.

I recommend black coffee, with a little cinnamon for even more antioxidants, and a scoop of hydrolyzed collagen for added skin-health benefits without changing the taste. If you absolutely must have a little milk, consider trying unsweetened almond, oat, or other plant-based milks to lighten your coffee. Also keep in mind that coffee is a mild diuretic, meaning it flushes out water. That can make your skin drier and more wrinkled, so drink a glass of water for every cup of coffee, and don't drink more than two or three cups a day—less if it makes you jittery.

What About Butter Coffee?

Butter coffee, more often known by the brand name Bulletproof coffee, contains one or two tablespoons of unsalted butter or ghee ideally from grass-fed cows, one to two tablespoons of MCT oil,* and one cup of good-quality coffee. Blend these all together for a creamy, frothy delight.

There are benefits to butter coffee. The grass-fed butter or ghee contains healthy omega-3 fatty acids, approximately 25% more than you will find in regular butter, and makes this coffee satiating enough to be its own breakfast. Meanwhile, the MCT oil can increase your metabolism to theoretically help you burn fat.[65]

---

* MCT oil is medium-chain triglyceride oil, which you can purchase in health food stores. Start slow with MCT oil, as it can take your digestive system some time to get used to it. Some people use coconut oil instead of MCT oil in their butter coffee. Coconut oil contains MCT oil, but is a less concentrated source.

But it is also high in saturated fat, which doesn't affect some people but isn't good for others, including those with heart disease. I consider butter coffee to be an occasional treat, especially on a cold winter morning.

- **Red wine.** Short of quitting smoking, one of the best things you can do for your health and appearance *might be* to have a small glass of red wine each day. As far back as ancient Greece, people have been extolling the health benefits of red wine, and it became trendy in the 1990s after a *60 Minutes* episode about the "French Paradox." Because people in France traditionally ate a lot of high-fat foods but had a low rate of heart disease, people began to wonder if wine consumption could explain this so-called paradox. When doctors confirmed the theory, the media didn't have to tell the masses twice that wine was healthy—people happily jumped on that bandwagon. Now we know more about why wine could have this effect: it's full of antioxidants, primarily the polyphenol resveratrol, which is most concentrated in red wine because it is most concentrated in the skin of grapes. (To make white wine, the grape skins are removed before they can color the wine, although white-wine fans may be interested to note that chardonnay has the most antioxidants of the white wines.)

  Resveratrol has been shown in laboratory studies to have anti-aging effects on animals,[66] but we still don't really have wide-scale scientific studies on resveratrol's effects on humans. Until we do, if you like wine and drink it regularly, then know that one small glass a day at dinner (i.e., with food) is plenty. Drinking more than this can actually make you feel and look older, due to dehydration and the toxic effects of alcohol, including worsening osteoporosis, diabetes, high blood pressure, stroke, ulcers, memory loss, and mood disorders.[67] That one small glass seems to be the sweet spot. For even greater antioxidant benefits, pair that glass of red wine with a piece of dark chocolate.

And if you don't drink wine? Or you don't like it? Or you know you have a problem with it and you can't stop at just one? Drinking red wine is certainly not necessary. There are plenty of other ways to get antioxidants. Have some actual grapes! Other foods with resveratrol include peanuts, cocoa, blueberries, and cranberries.[68]

Organic Red Wine

Full disclosure: I'm not a big wine drinker. Although I know a glass a day may be great for my health, my job and lifestyle don't allow me to partake as often as I'd like. So, although I don't drink it very often, when I do, I really like Dry Farm Wines. These are the purest, highest-quality artisan wines that are free of additives like sugar, pesticides, herbicides, artificial dyes, and commercial yeasts. This wine is organic, and it's the only wine that doesn't make my face flush or give me a hangover the next day. A lot of people in the holistic health field choose this company, and I recommend them for anyone who enjoys wine.

AUTOJUVENATING FOOD LIST: ANTIOXIDANT FOODS

These foods are all high in antioxidants, and many are also anti-inflammatory. Eat as many of these foods as you can. Please visit autojuvenation.com to download a free printable shopping list of these and other autojuvenating foods, that you can bring to the grocery store.

FOODS HIGH IN VITAMIN C

Blackberries
Blueberries
Broccoli
Brussels sprouts
Cabbage
Cantaloupe
Cauliflower
Gooseberries
Grapefruit
Kale
Kiwi fruit
Mangoes

Melon
Oranges
Papaya
Pineapple
Pomegranates
Raspberries
Red bell peppers
Rhubarb
Strawberries
Sweet potatoes
Tangerines
Tomatoes

FOODS HIGH IN VITAMIN E

Almonds
Asparagus
Avocados
Beet greens
Broccoli
Butternut squash
Carrots
Collard greens

Dried apricots
Mangoes
Peanuts
Pumpkin
Spinach
Sunflower seeds
Wheat germ oil

FOODS HIGH IN CAROTENOIDS

Arugula
Broccoli rabe
Brussels sprouts
Cabbage
Cantaloupe
Carrots
Collard greens
Corn

Grapes
Kale
Kiwi fruit
Orange or yellow
  peppers
Parsley
Romaine lettuce
Spinach

Squash   Turnip greens
Sweet potatoes

OTHER ANTIOXIDANT-RICH FOODS

Coffee
Dark chocolate
Green tea
Red wine

CHAPTER EIGHT:

AGE-ACCELERATING FOODS TO AVOID

I like to focus on what we *can* do to autojuvenate, but when it comes to food, it's also important to think about what we might not want to do anymore, if we don't want what we eat to cause premature aging. Many foods are high in calories and low in nutrients (in other words, *not* nourishing). Some foods are actively damaging to health, such as those that are inflammatory, degrade collagen, and generate excessive free radicals. While we're all likely to eat these foods once in a while (I include myself, so don't call me out if you see me eating any of these foods in public!), these foods should ideally not be a daily part of any diet focused on autojuvenation.

HOW FOODS CONTRIBUTE TO AGING

Remember when I told you that there are five major processes that account for the majority of the aging of our bodies and our skin? As a refresher, they are:

1. Nutrient depletion
2. Inflammation
3. Collagen degradation
4. Oxidative stress (free radicals)
5. Declining autophagy (resulting in a buildup of cellular waste)

The foods that contribute to these aging processes are the foods to avoid most of the time because they can age your insides, degrade your skin, and even possibly shorten your life. They come in a few different categories, each of which contributes to aging in multiple ways.

Some foods are literally AGE-ing, meaning they can trigger advanced glycation end products, or the AGEs I discussed earlier. To put it simply, AGEs result when sugar reacts with proteins or fats in a way that compromises both the structure and function of a healthy cell.[69] The sugar molecules attach to the collagen and elastin in our skin, causing them to become deformed. This can result in wrinkles and sagging skin.

But it's not just sugar. Animal foods high in fat and protein naturally contain large amounts of AGEs[70] which increase with cooking. In fact, any foods cooked at high heat, such as with roasting and especially frying, also contain a lot of AGEs. So do certain processed foods such as breakfast cereal.[71] The foods that contain the most AGEs are red meat, high-fat cheeses, butter, margarine, oils, roasted nuts, egg yolks, and fried food in general, although any high-temperature cooking, including grilling and broiling, increases the AGE content of foods.[72] The food with the highest amount of AGEs? Bacon (which contains both sugar and fat). Yes, you did just see a tear slide down my cheek.

These complex products age you, inside and out. They cause chronic inflammation in internal organs and on the skin and have been implicated in everything from premature wrinkles to diabetes, heart disease, cataracts, and dementia.[73] In Marvel movie terms, AGEs are like Thanos and his intergalactic minions coming to destroy earth (your body).

In addition to containing AGEs, some foods contribute to the generation of free radicals. These are generally foods that are highly processed so that they become, inside the body, almost like toxic chemicals, compromising healthy function by generating destructive reactive oxygen species.

These include fats and oils that have been oxidized—that are rancid or that have been heated multiple times for frying—as well as processed meats (like bacon, sausage, and deli meats) that are chemically preserved. Excess sugar in the blood, as is common in diabetes or prediabetes as well as in people who eat a diet heavy in highly refined grains and sugars, can also generate excessive free radicals.[74] In addition, alcohol can generate free radicals as it gets broken down in the liver. Plus, it stunts the body's ability to generate and use antioxidants.

Inflammation itself is another way food can cause premature aging. The most inflammatory foods are fried foods, foods with a lot of processed sugar (especially sugar-sweetened beverages), refined grains (like white flour and white rice), cured meats like bacon, salami, hot dogs, and ham; and foods that are highly processed.[75] Are you noticing some repeat offenders?

Foods with a lot of omega-6 fatty acids can also be inflammatory. We need omega-6 fatty acids to help with the process of inflammation (less inflammatory sources include walnuts, peanuts, sunflower seeds, hemp seeds, tofu, avocado oil, and eggs), but they need to be balanced by omega-3 fatty acids to resolve the inflammation. The ratio of omega-6 to omega-3 is the key. Today's standard American diet is overly rich in omega-6s and low in omega-3s. This ratio needs to move closer to 1:1 for best health. You can do this by eating more fatty fish and other omega-3-rich foods, and eating fewer foods rich in omega-6, especially those containing a lot of vegetable oil, like sunflower, corn, soybean, and canola oil.

Beyond what you eat is how often you eat. If you eat all day, especially a lot of protein and carbs, your body won't get that down time for cellular cleanup, and that, too, promotes premature aging.

Different types of foods can age you in different ways, so let's look more closely at what's going on when you eat certain categories of age-accelerating foods.

REFINED CARBS AND SUGAR

Refined carbs and sugar increase the process of aging through glycation (via the creation of AGEs) and by increasing inflammation. Yet, more than 40% of the calories in the American diet come from sugar and refined grains, and for many people, that percentage is much higher. The average American consumes 152 pounds of sugar every year, and about 20% of our daily calories come from sugar-sweetened beverages alone—soda (which we call *pop* in Michigan), sports drinks, sweet teas, sweetened coffee drinks, and juice.[76] One study attributed 184,000 deaths every year to the effects of drinking sugary drinks, which have been shown to cause or contribute to obesity, heart disease, cancer, and type 2 diabetes.[77]

Sugar and refined grains are largely void of nutrients and can therefore contribute to malnourishment. Much of the natural vitamin and mineral content that comes with whole grains is removed, leaving only the starch and sugar, so your body may naturally crave more food to make up for the missing nutrients. This is probably one of the reasons why sugar and refined grains make people even hungrier. That can lead to overeating and excess body-fat accumulation, which is in itself inflammatory.

We don't technically need to eat any sugar or refined grains, but our bodies do need sugar—we make it primarily from foods containing carbohydrates. Pretty much everything we eat gets broken down into sugar or, more accurately, glucose, and glucose is the body's preferred fuel source. Glucose and insulin work together in order to transform food into energy. Every time you eat, your pancreas secretes insulin, which sends glucose into cells. The cells can then use glucose for energy. The rate at which this happens, and how high or low your blood sugar and insulin go, depend on what and how much you eat.

If you eat foods with a very high glycemic index (a measure of how quickly a food raises blood sugar on average), your blood sugar can go very high, resulting in a spike in insulin, which can then result in low blood sugar and blood-sugar instability overall, when excess insulin drives too much sugar into the cells. When this happens too often over time, the cells of your body (such as your muscle and fat cells) can become resistant to the effects of insulin. They begin to lock glucose out of the cell, leaving it to circulate in the bloodstream. This drives up blood sugar, which can lead to prediabetes and, eventually, type 2 diabetes. The same phenomenon can happen when you eat too much. Your blood sugar will surge, and then your insulin will surge. This is what people mean when they talk about the blood-sugar roller-coaster.

To keep blood sugar and insulin levels steady, eat smaller portions and more foods with a low glycemic index. This is how the body is supposed to work. A healthy body can typically handle the occasional high blood-sugar spike and insulin surge, but problems arise when these spikes happen too frequently. They wear down our cells and result in premature aging.

There are other negative effects of high blood sugar, besides an unstable glucose/insulin balance. Even before your body's cells become resistant to the effects of insulin, high blood sugar can have a noticeably adverse effect on skin, making it more oily and acne-prone. For example, one study found that a low-glycemic diet was correlated with a lower incidence of acne in young men.[78]

High blood sugar can also cause glycation, which is the process I described earlier, when glucose (and fructose) bind to collagen and elastin, damaging them and resulting in a loss of tightness and elasticity.[79] That doughnut can literally make your skin sag, and these aging effects can become noticeable by the age of 35, then continue to worsen with increasing age if you keep eating those doughnuts (or other junk food).[80] To put it in clear terms, a study in *The American Journal of Clinical Nutrition* found that people who eat more carbs experience more advanced aging.[81] (Just a reminder that refined carbs and sugar are usually the problematic carbs—complex carbs like whole grains and legumes in moderation are worth eating for the microbiome-boosting effects.)

Spiking insulin causes problems, too. For one thing, it can result in increased oil production and increased testosterone, especially in women, which can lead to acne breakouts. This is one reason why women with PCOS (polycystic ovary syndrome), who tend to have blood-sugar issues, also tend to have higher testosterone and problems with acne.

Another downside of excess insulin is chronic inflammation, which can have aging effects both inside and out. In addition to its link with chronic disease, chronic inflammation can worsen rosacea and rashes, increase oiliness, and weaken the collagen and elastin in the skin.

So what can you do to keep your blood sugar and insulin under control? My best advice is to limit processed foods and foods with a high glycemic index, especially refined simple carbohydrates like white bread, white potatoes, white rice, baked goods made from white flour, and foods with added sugar.

Fructose: The Reason Some People Think Fruit is "Bad"

Is fruit bad for you? Absolutely not. Fruit is one of the healthiest, most nutrient-dense foods. However, fruit contains a natural sugar called fructose that tends to have an adverse effect on blood sugar. You probably already know that high-fructose corn syrup, a highly processed and ultra-refined sweetener, is terrible for your health and blood-sugar stability, but some people will also get blood-sugar spikes from whole fresh fruit that is high in fructose, like grapes, pineapples, watermelon, overripe bananas, dried fruit, and canned fruits. Lower-fructose fruits include berries, apples, and citrus fruits (although oranges are the highest in fructose of the citrus fruits), so these may be better choices for people who have problems with blood-sugar stability. For healthy people, the old saying is true! An apple a day could actually keep the doctor away. For people with prediabetes, type 2 diabetes, or who have a family history of these problems, blueberries and oranges may be even better choices.

WHAT ABOUT GLUTEN AND OTHER GRAINS?

When it comes to gluten specifically and grains in general, there are two camps. One says grain isn't good for you and gluten-filled grains (like wheat, barley, rye, and spelt) are inflammatory for everyone, so nobody should eat gluten and everyone should at least limit grains, including whole grains. The other camp says that grains are an excellent source of fiber and minerals, that whole grains are good for you, and that unless you have celiac disease, gluten is a fine source of protein, and whole grains are the staff of life.

So what's a health-conscious person who seeks autojuvenation supposed to do about (to use just one example) delicious bread? What about pasta? Rice? Is quinoa bad? Can you never again enjoy a cookie or a slice of cake?

As with most things, the truth is probably somewhere between the two extremes, and what is best for you might be different than what is best for someone else.

Studies have shown that whole grains can be an integral part of a heart-healthy diet.[82] However, if you are sensitive to the negative effects of gluten, then nongluten grains may be a less inflammatory choice for you. The problem is that many people don't know they are sensitive to gluten. They may have a lot of digestive or skin issues that they don't realize are aggravated by gluten or, worse, chronic pain and joint inflammation that is related to gluten sensitivity.

One compelling argument against gluten is that people with gluten sensitivity, whether celiac disease or otherwise, have higher levels of zonulin in the blood. Zonulin is a protein that regulates the integrity of your intestinal barrier,[83] the epithelial lining of the intestines. Zonulin has been associated with leaky gut syndrome, in which food proteins, bacteria, and other particles penetrate (or "leak") through the intestinal lining and cross into the bloodstream. It's believed that this can confuse the immune system, which attacks the proteins as pathogens, creating an unnecessary or inappropriate autoimmune response.

Some researchers believe that gliadin, one of the proteins in gluten, stimulates the overproduction of zonulin, resulting in leaky gut. This may be why people with autoimmune diseases (such as rheumatoid arthritis, lupus, or multiple sclerosis) may do better on a gluten-free diet. Leaky gut has also been associated with inflammatory skin disorders[84]—another case of the skin being the magic mirror that reveals what's going on inside.

But studies show only about 1 to 3% of the population has celiac disease. A much higher percentage might be sensitive to the effects of gluten, but only you can know if you are one of these people. Do you get digestive upset, itching, headaches, brain fog, or discomfort after eating gluten? If you aren't sure, it might be worth removing all gluten, or even all grain, for a couple of weeks (ideally four to six weeks, although two weeks may be enough to give you an idea). Notice if your symptoms subside. If they remain the same, the problem probably isn't grain or gluten. If they lessen or go away, you may be on to something. After your system has calmed down, you can try reintroducing gluten-containing or grain-containing foods, to see if you get a recurrence of your symptoms. (Technically, this is called an elimination diet.)

Some people may be reactive to large amounts of gluten, but not to small amounts. Some researchers believe gluten intolerance is a spectrum, with people who have true celiac disease on one end (they should never ingest any gluten) to people who have no reaction at all on the other. You may fall anywhere between those extremes, doing okay with small amounts of gluten but having symptoms if you eat large amounts or eat it every day.

I admit it's very difficult to avoid gluten, and grains in general even more so. Most of us enjoy a good a slice of bread now and then, or regular pasta, or other grains like rice. My general advice is to try going without gluten for a while to see if you feel better. Many of my patients say they feel much better by removing or even just reducing gluten in their diets. I've had patients whose joint pain and swelling went away, or whose IBS symptoms improved dramatically. Some even saw their skin inflammation clear up. Find out for yourself what works for you.

And if you are a die-hard grains fan, and you don't think grains or gluten bother you, you can still make superior nutritional and anti-aging choices when it comes to grain products.

EATING GRAINS FOR AUTOJUVENATION

First of all, if you have celiac disease, don't eat any gluten-containing products, and if you think grains upset your stomach or may cause you to feel crummy, take a break and see if that helps. However, if you do want to keep grains in your life, here are my best tips for including grain in the best way possible, with the most fiber and nutrients.

- **Choose sprouted grain** over flour-based breads. Sprouted grain breads are made using ancient grains that haven't been hybridized over the years, as well as legumes and seeds. They don't use flour, which is highly processed. (Ezekiel is one good brand.) Sprouted grain breads are generally devoid of added sugar, preservatives, and artificial ingredients, and they have more protein, fiber, vitamins, and minerals, and in some cases less gluten, than traditional bread, making them easier to digest. If you are going to eat bread, I highly recommend choosing sprouted grain breads made with ancient grains, for the healthiest option that is the easiest to digest.

- **Avoid refined grains** or grains stripped of their nutrients and fiber, made into white flour, including most baked goods, pasta, and white rice. Refining grains removes the bran and germ, which is where most of the fiber and essential nutrients are. While white flour leads to a lighter product that tastes less heavy and bulky, it also digests more quickly, causing a bigger spike of glucose and subsequently a bigger spike of insulin. I recommend limiting anything made with white flour—white flour is not good for your skin, causing inflammation and glycation (AGEs). Think of eating white-flour foods the same way you think about eating sugar.

AUTOJUVENATING FOOD LIST:
LOW GLYCEMIC FOODS[85]

GRAINS

Barley	Rye
Bran cereals	Whole wheat pasta
Bulgur wheat	Whole wheat tortillas
Oatmeal (rolled oats or steel cut, not instant)	Wild rice

FRUITS

Apples
Apricots
Avocados
Blackberries
Blueberries
Cherries
Cranberries
Grapefruit

Olives
Peaches
Pears
Plums
Raspberries
Strawberries
Tangerines

BEANS AND LEGUMES

Black-eyed peas
Butter beans
Chickpeas (garbanzo
   beans)
Green beans
Hummus
Kidney beans
Lentils

Lima beans
Navy beans
Peanuts
Pinto beans
Snow peas
Soy products (like tofu
   and tempeh, organic
   if possible)

VEGETABLES

Artichokes
Asparagus
Broccoli
Cabbage
Cauliflower
Celery
Cucumber
Eggplant
Leafy greens
Lettuce

Mushrooms
Okra
Onions
Peppers
Spinach
Summer squash
Tomatoes
Turnips
Zucchini

NUTS AND SEEDS

Almond milk (make
sure there's no added
sugar)
Almonds
Hazelnuts

Pecans
Pumpkin seeds
Sunflower seeds
Walnuts

EGGS, MEAT, AND SEAFOOD

Chicken and turkey
Eggs
Fish, all types

Red meat
Shellfish

AGE-ACCELERATING FOOD LIST:
HIGH GLYCEMIC FOODS TO AVOID[86]

BREAKFAST CEREALS

(most are not primarily fiber-based)

BEVERAGES

Fruit juices
Soda

Sweet tea

GRAINS

Cornmeal
Couscous
Pasta, white

White bread, rolls,
bagels, tortillas,
muffins, and most
baked goods

SNACK FOODS

Cakes
Candy

Chips
Cookies

Crackers
Doughnuts
Granola bars
Jams and jellies
Pastries

Pretzels
Rice crackers
Syrup
Tortilla chips

BAD FATS

Some fats, as you've already learned, are good for you, but others are not. Worst on the list are artificial trans fats, which are industrially modified plant oils that are cheap and don't spoil quickly, so they are widely used in food processing, packaged foods, and fast foods. Recognizing how dangerous to health trans fats are, the FDA has banned them, but they still show up in trace amounts in processed foods, so it's a good idea to look at the ingredients list on any packaged foods. If you see *partially hydrogenated* anything, back away! The medical community is in agreement that trans fats contribute to cardiovascular disease, breast cancer, premature births, colon cancer, diabetes, obesity, and allergies.[87]

But trans fats aren't the only fats that contribute to aging. I've mentioned the benefits of omega-3 fatty acids already, and also mentioned that omega-6 fatty acids aren't good for health when consumed out of proportion. Omega-6 fatty acids are found in large amounts in processed food, refined vegetable oils (like corn, soybean, and safflower), and conventionally raised meat. Too much omega-6 has been shown to depress immune function, contribute to weight gain, and generally cause inflammation.[88]

Instead of using vegetable oils, stick to olive and avocado oils (technically these are fruit oils), which are much less processed and without exposure to the solvents, bleaching, and deodorizing that is common in more highly refined oils. Both olive and avocado oils have a much better ratio of omega-3 to omega-6 fatty acids.[89]

Age-Accelerating Foods: Soybean Oil Alert

Soybean oil constitutes over half of all the oil we consume in the US. It's ubiquitous in processed foods, fried foods, salad dressings, and baked goods, and it has high levels of omega-6 fatty acids. Most soybean oil is genetically modified, and it's also highly processed and therefore unstable at high heat, quickly becoming rancid. Although it's virtually impossible to avoid eating soybean oil completely in today's world, try to limit your intake.

The last type of questionable fat is saturated fat. This is still a matter of debate, with reputable experts on both sides of the question asking, *Is saturated fat good or bad for you?* We used to believe that too much saturated fat contributed to heart disease and inflammation, and some evidence still points to this. For example, a study in the *Journal of the American College of Nutrition* found that people who ate a lot of butter and margarine—both high in saturated fat—had more skin aging and wrinkling than those who ate less (the implication being that inflammation causes premature skin aging).[90] However, this is an older study from 2001, and it's not clear that saturated fat was the actual culprit.

We do know that saturated fat consumed along with foods that are high in refined carbs and low in omega-3 fatty acids can cause inflammation,[91] but when combined with lower-carb, higher-fiber foods rich in omega-3 fatty acids, saturated fats don't seem to have as much of an effect on inflammation.[92] It may be the combination of refined carbs and saturated fat so common in the standard American diet that is harmful, rather than the saturated fat itself.

When you consider that most food consumed in the industrialized world is high in refined carbs and low in omega-3 fatty acids, saturated fat may indeed be harmful for a majority of the population. For this reason, I recommend limiting your intake of saturated fats. If you do often choose foods high in saturated fats, I recommend they come from real, unprocessed, or non-factory-farmed foods, and that you avoid eating them with refined carbs. This strategy will likely be much better for your skin and overall health.

What About Cholesterol?

Does dietary cholesterol translate to blood cholesterol? Scientists used to say yes, but now the prevailing opinion is that dietary cholesterol (such as in eggs) does not contribute to blood cholesterol—at least, not directly. Many foods (such as eggs) that contain dietary cholesterol also contain saturated fat, too much of which may be linked to blood cholesterol. The Dietary Guidelines Advisory Committee and the American College of Cardiology/American Heart Association removed its advice to restrict dietary cholesterol,[93] stating that cholesterol should no longer be considered a nutrient of concern for overconsumption. Others disagree, so I would say the jury is still out on this one. Moderation is likely the best practice here. Have your eggs...but maybe not more than one or two a day.

ULTRA-PROCESSED FOOD

This one is surely no surprise at this point. Think of processed food as fake food. Putting fake food into your body creates all sorts of reactions that can prematurely age you and harm your health.

Technically, processed food is food that has been altered from its natural form, and ultra-processed food, or UPF, has been extremely altered through chemical processes with many added ingredients for coloring, flavoring, and preserving. For example, an apple is a whole food. Applesauce is a minimally processed food. A deep-fat-fried, sugar-loaded, packaged or fast-food apple pie is an ultra-processed food. You can check the ingredients list for added chemicals, like preservatives, artificial colors and flavors, dough conditioners, etc. to get an idea of how processed something is, but you can also tell just by looking. Did the food come that way naturally for the most part, or would it be unrecognizable to someone from 100 years ago?

But maybe you want more specific reasons why you should give up your favorite salty or sweet snacks. Here are my top seven reasons why processed food is age-accelerating:

1. **Free radicals.** Processed foods often contain (and are often fried in) refined vegetable oils like soybean oil which are highly oxidative and contribute to free-radical formation, especially when heated.[94] These foods usually don't contain antioxidants that can help to mitigate this free-radical assault.

2. **High-fructose corn syrup.** This is the worst type of sweetener to ingest. It results in glycation and inflammation.

3. **Refined carbs.** Almost all processed foods contain refined carbs, which can result in sugar spikes, insulin spikes, and inflammation. Many also contain gluten, which can be inflammatory for many people and dangerous for some.

4. **Preservatives and other chemicals added to extend shelf life.** In order to keep them supposedly fresh for a long time without refrigeration, many processed foods are loaded with chemicals produced in labs and factories. This is one significant reason why processed food should be considered fake. Like journalist Michael Pollan famously said, "If it came from a plant, eat it; if it was made in a plant, don't."

5. **GMO ingredients.** Processed foods usually contain some

combination of corn and/or soybeans, both of which are overwhelmingly genetically modified in the US food supply. The concern with GMO foods is twofold. Not only do we not yet know what kind of health effects they have, but also, these foods are modified genetically to withstand higher doses of potentially harmful pesticides and herbicides, such as glyphosate, the herbicide in the ubiquitous agricultural product called Roundup. These chemicals are widely used in industrial farming, and are therefore prevalent in processed food. When you see "GMO" on a label, it's a good bet that the food is high in pesticide or herbicide residue—most likely glyphosate. In fact, a study performed by the Environmental Working Group (EWG) found high levels of glyphosate in several popular children's cereals![95]

6. **Addictive.** Food scientists engineer processed foods to be what they call "hyperpalatable," meaning they overstimulate our taste buds and the pleasure centers in our brains in a way that is nearly indistinguishable from what addictive drugs do. They taste so good that they trigger overconsumption. The purpose is to increase profits for giant food corporations, even though the food scientists themselves know that these foods are harmful to health. Foods made by nature don't have this effect. Compare potato chips and strawberry slices. Both are delicious, but which one is harder to stop eating?

7. **Low in fiber.** Processed foods often have the fiber stripped away to make them taste lighter and feel less filling, which also leads to overconsumption. Fiber is an important prebiotic for your beneficial gut bacteria, and fiber-poor foods can contribute to the overgrowth of more pathogenic bacteria. Foods with little fiber are also more likely to trigger blood-sugar and insulin spikes. Short version: processed food is bad for your microbiome health and your blood-sugar balance.

THE DAIRY DILEMMA

Are you hoping I won't say anything bad about cheese? Well, get ready. Dairy products are incredibly popular but I think by any measure, we eat too much of them. For one thing, the average American consumes 630 pounds of dairy products per year![96] And while there is some research that dairy products can be a good source of nutrition, they also contain a lot of sugar (in the form of lactose, which many people are intolerant of) and a potentially inflammatory protein called casein. While not everyone is sensitive to lactose and/or casein, many people are, especially if they are not of Northern European descent. In fact, 95% of Asians, 60 to 80% of African Americans, and 50 to 80% of Latinos are lactose-intolerant. Only 2% of Northern Europeans are. Worldwide, most people are lactose-intolerant, myself included.

During high school and early college, I had some digestive issues. They worsened considerably late in college, and I figured out that they were at least partially due to lactose. The kicker came one day when I decided to eat an entire Wendy's Frosty. I grew up eating and loving Frostys, but I quickly found out that they no longer loved me. After spending two hours on a toilet praying for death to end my suffering, I decided to severely curtail my dairy intake. I haven't eaten a Frosty since.

But lactose and casein aren't the only potential problems with dairy products. There are other reasons why I don't recommend dairy intake as part of an autojuvenating diet. A lot of research on dairy comes out of Sweden, where its consumption is quite high. A 2021 study from Sweden published in *Ageing Research Reviews* found that the more pasteurized milk people drank, the more they were likely to die (from any cause) during the course of the study. The researchers concluded that the reason for this was likely related to a complex biochemical process.

The branched-chain amino acids, lactose, and microRNAs in dairy products enhance signaling of something called mTORC1, which is linked with aging and age-related disease.[97]

It's beyond the scope of this book to get deep into these mechanisms, but suffice it to say that too much mTOR from dairy products is aging. Intriguingly, the study also found that fermented milk products (like yogurt) contained microbes that blocked this effect and were therefore not associated with increased mortality and aging. So if you love your yogurt, it may not be harmful to you, especially if you don't eat the kind with added sugar. And if you're going to eat dairy products, yogurt may be the best form.

But there are still other reasons to avoid dairy. Hormones are a problem with commercial milk, which contains large amounts of both estrogen and progesterone, with one study revealing, "Sexual maturation of prepubertal children could be affected by the ordinary intake of cow milk."[98] Dairy products also contain antibiotic residues, just as factory-farmed meat does.[99] Finally, the protein in milk is about 20% whey and 80% casein. Casein's inflammatory properties have been demonstrated in studies to increase the risk of cancer by stimulating the production of insulin-like growth factor 1 (IGF-1).[100]

SALTY FOODS

A little salt isn't going to hurt anyone, but too much salt is a disaster for skin as well as aging. Salty foods cause water retention, which results in puffiness and bloating. Have you ever noticed that you have puffy eyes the morning after eating a dinner of Chinese takeout? That's because of the salt and the MSG (monosodium glutamate, which contains sodium). I hate to break it to you salt fans, but chronic puffing of the eyelids can stretch out the skin, making it look more wrinkled and aged.

Remember telomeres? Excess salt has also been shown to

shorten telomere length in people who are overweight, and that is a key indicator of cellular aging.[101] Shorter telomeres are associated with accelerated aging and a shorter life span (see page 249).

SUGARY DRINKS

This isn't the first time I've mentioned the hazards of sugar-sweetened beverages, but I'll add that regular consumption of soda pop also results in shorter telomeres.[102] It doesn't take much. Drinking just eight ounces of soda per day (less than a can) corresponded, in the study I've referenced above, to 1.9 years of accelerated aging. Drinking a daily 20-ounce serving (one of those bottles labeled *single serving*) resulted in 4.7 more years of aging and is as damaging to longevity (at least, in terms of telomeres) as smoking.

But don't think diet soda is the answer to your soda cravings and anti-aging goals. It's associated with weight gain, possibly because of how it affects our taste buds.[103] Even without calories, diet soda still tastes sweet and so it encourages habituation to sweet tastes. This can then lead to eating more sweet foods, such as candy and frosted desserts, in general. That can lead to blood sugar issues as well as weight gain, which can in turn lead to lower energy and more health problems. There is also compelling evidence that artificial sweeteners affect the microbiome in ways that impair healthy blood sugar responses.[104]

Now that we know how hazardous regular and diet soda pop consumption can be, I hope we can steer children away from it. I certainly grew up drinking all types of soda pop, including A&W Root Beer, Coke, 7 Up, Fanta Orange, and my least favorite, Hubba Bubba bubblegum-flavored soda. I have no idea why my mom bought two cases of Hubba Bubba soda one day, but it sat in our house for months—which was probably best for all involved!

Although the best thing to do is to just not drink soda pop at all, if you are like the millions of Americans who are addicted to drinking it each day, you can start by reducing your consumption. Dropping from three daily cans of soda pop to two or one is a big success! The goal would be to eventually not drink it at all or simply consider it a rare treat. Believe me, your body will thank you for it! I absolutely love the taste of an old-fashioned Coke in a glass bottle and will admittedly indulge in one on special occasions, but I no longer drink soda pop regularly.

BURNT AND CHARRED FOODS

What's wrong with burnt toast or burnt bacon? The problem with burning and charring sounds like a chemistry lesson, and in a sense it is. Burning or charring foods, especially meat on the grill or bacon on the griddle, causes the formation of heterocyclic amines, which have been shown to cause cancer.[105] These compounds also create AGEs, those proinflammatory compounds I keep mentioning caused by glycation that can speed up the aging process. Is that charred steak really worth increasing your risk of cancer and making you look older? I don't think so. The best cooking techniques for minimizing dietary sources of AGEs and heterocyclic amines are boiling, steaming, slow-cooking, and low- to medium-heat baking. And of course, eating your food raw. I enjoy an old-fashioned summertime cookout as much as the next plastic surgeon, but I encourage you to try to avoid charring the meat, for reasons of taste *and* your health.[106]

AUTOJUVENATION FOOD LIST

To avoid AGEs, non-nourishing calories, inflammatory foods, collagen-wrecking foods, and free-radical-generating foods, avoid the foods on this list most of the time. I'm not saying you can't ever eat cake or french fries again, but if you are serious about autojuvenation, you will want to avoid eating these foods regularly.

- Refined carbs and sugar
- "Bad" fats
- Processed food
- Conventionally produced meat
- Dairy
- GMO foods
- Excess salt
- Sugar-sweetened and artificially sweetened beverages
- Burnt and charred food

Bottom line, every bite of food you eat could potentially do at least one of four things: nourish you, cool inflammation, promote collagen production, or reduce oxidative stress by neutralizing free radicals. Many foods check multiple boxes.

Every time you choose to eat something, pause and ask yourself, *Is this working for me or against me? Will this accelerate aging or promote autojuvenation?* Thinking about food in these terms can help you to see your dietary choices in a completely different light. Those so-called pleasurable foods might begin to seem more insidious, and those healthy foods you always forget to eat might suddenly seem indispensable.

That being said, I also understand that we only have one chance to live on this planet, and there is definite joy in having a slice of cake at your daughter's birthday party, a greasy burger with your friends, or a scoop of gelato while on vacation. A tasty splurge is totally fine every once in a while, even if it's not anti-aging. Life is always better when we can enjoy it!

And if I see you eating a cheeseburger and crinkle fries at your local Shake Shack, don't feel guilty. I'll probably be eating them, too.

118

Just not all the time.

To help you integrate more autojuvenating foods into your diet without having to flip around to different chapters, I've compiled a list of all the foods mentioned in this chapter and elsewhere in this book, that you can find in Appendix 1. I hope you will consult it whenever you are trying to figure out what to eat. I've also created a printable shopping list of these foods which you can download for free at autojuvenation.com. Feel free to bring it to the grocery store with you as reference.

CHAPTER NINE:

STOP EATING TO PROMOTE CELLULAR REJUVENATION

The last few chapters have been all about the amazing things you can eat to slow down the aging process and make your outside and insides look and feel younger. But there is a second part to the dietary part of the Younger for Life Program. In our Autojuvenation Jump Start, you'll see that feasting (what to eat) is Phase 1, but Phase 2 is about fasting.

Feasting *and* fasting can impact aging. While feasting on autojuvenating foods is crucial, fasting (and some of its variations, like intermittent fasting) gives your body a chance to do another very important anti-aging task: clean house. This process is called autophagy, which means self-eating, and as I've briefly alluded to in previous chapters, this is a fundamental cellular cleanup process that allows our cells to stay young and healthy by making way for new cells. When autophagy kicks in, your body cleanses cells by removing aged, damaged components (organelles, proteins, and cell membranes). Your body does this automatically—unless you are eating all day long. When you give your body periodic breaks from constant digestion, it can switch into autophagy mode. Instead of incorporating nutrients, it can take a break and take out the trash.

I practice intermittent fasting a couple of times a week, but not every day. Some people do practice it daily and swear by it, but I find that what I do is enough and works for me. When I'm in the operating room, I function best with a high-protein breakfast. However, on the mornings when I'm not going to be in the OR, I will often fast.

My first real experience with fasting began several years ago. I tried something called the ProLon Fasting Mimicking Diet. This is a five-day, limited-calorie fast that tricks your body into thinking you're actually completely fasting. It was developed by Dr. Valter Longo, the director of the Longevity Institute at the University of Southern California.

I wanted to try this because although I was eating a healthy diet, I had gained a few unwanted pounds. Having read a little bit about it, I decided to try it out. The diet is based on eating a certain amount of specific food each day, carefully chosen by Dr. Longo for its ability to (essentially) trick your body into thinking you aren't eating when you actually are. The purported benefits are weight loss, autophagy, and stem-cell creation.

On the first day, I ate a total of 1100 calories. Not much, but doable. On days two through five, however, I decreased my calories (according to the instructions) to between 700 and 800 calories. I had never fasted before, or even intermittently fasted, so this was a new thing for me. I was a little hungry on day one but had good energy, and overall, I felt pretty good. Days two and three were tough, though. I was really hungry and felt weak and even light-headed at times. I wasn't sure I could keep going. But by day four, I started to feel a little better—even though I admit I was side-eyeing that slice of pizza my daughter was eating. I remember thinking, *Hey, you've done four days, you've lost a couple of pounds, you know what it's like to do this, why not just stop now?* But I persevered through the last day, and that's when something very odd happened.

On day five, I woke up feeling refreshed, clearheaded, and energetic. I felt *great*. I was eating very little, but my body had stopped craving food. According to Dr. Longo, I was entering a period where my body was in full autophagy mode, which meant it was recycling damaged and aged intracellular organelles for energy and gearing up to create fresh stem cells. Honestly, it was bizarre how amazing I felt and how good my skin looked: it was smoother and more radiant than it had been in months.

Overall, I lost seven pounds doing the Fasting Mimicking Diet. I hadn't weighed that little since college! I also felt like my microbiome had been reset. During the diet, I had some discomfort with (TMI alert) bowel movements, maybe because there wasn't much in there to move. Afterward, my bowel movements were more regular and formed than they had been in months. The pounds I lost stayed off, and the experience taught me something super important: I don't need to eat as much as I had been eating before I tried the diet. This also got me into exploring intermittent fasting as a way to continue what I had started. Since then, I've repeated the Fasting Mimicking Diet once, but otherwise, I reap the benefits of fasting through intermittent fasting instead.

THE TRANSFORMATIVE EFFECTS OF AUTOPHAGY

Regularly getting into a state of autophagy is a key for increasing longevity and health span, as well as slowing the aging process. Imagine you're remodeling your kitchen. There are two steps. First, the demolition, when you break down and remove the old things that don't work anymore. Then, the installation, when you put in the new floor, the new appliances, the new backsplash. If you don't get the old stuff out, you won't have room to put the new stuff in. Autophagy is the demolition that allows you to fill your cells with fresh new material, enabling everything to work better...to work younger.

Fasting, as well as eating certain foods and restricting calories, are the primary ways to trigger this process. When your cellular functioning stays young, so do you. Scientists have been studying this process with promising results. One study showed that autophagy promotes longevity in our bodies by removing toxic materials from our cells and recycling them as an alternative nutrient source.[107] The body wastes nothing it can use.

Our bodies are meant to undergo autophagy on a daily basis, but if autophagy isn't activated regularly, our cells can fill up with cellular waste, impeding function and degrading cells. This is what can result in accelerated aging, both inside and out. Unfortunately, autophagy slows with age, which is one likely reason why people who are older tend to feel more run-down and are more disease-prone.[108] As we age it's important that we create more opportunities for autophagy. Let's look in more depth at the most effective strategies for activating autophagy.

CALORIE RESTRICTION

The first time people got excited about calorie restriction was in 1917, when a study showed restricting calories in rats significantly increased their life span.[109] It sounds easy enough: calorie restriction for health purposes typically involves cutting calories by 20 to 30%. For example, if you currently eat a 2,000-calorie diet (a typical amount in the US), to practice calorie restriction you would cut your calories down to 1400 to 1600 per day. Some people who practice more extreme calorie restriction will cut down by 40%, which would be 1200 calories for someone on that 2000-calorie diet.

But even cutting your calorie consumption by 15%—that's taking our 2000 calorie diet down to 1700 calories—can, according to a recent study on healthy, young people without obesity, significantly reduce the production of reactive oxygen species,[110] or free radicals. This could in turn reduce chronic inflammation and slow accelerated aging.

Not only does calorie restriction reduce free-radical production, but researchers think it may also significantly increase time for autophagy.[111] The less digesting your body has to do, the more time it can spend on cellular cleanup.

Calorie restriction is now well-established in the scientific community as a way to increase life span and even reduce the risk of chronic diseases, not only by decreasing free-radical production but by affecting metabolic and hormonal factors that have been implicated in the disease processes that lead to diabetes, heart disease, and some cancers.[112] One study concluded that autophagy via calorie restriction could be an effective way to increase longevity because of the way it removes damaged organelles and proteins, improving cellular functioning.[113]

But there is a problem with calorie restriction, and it's not an insignificant one. In our current culture, it's extremely difficult to maintain calorie restriction for long periods of time. Our bodies have become comfortable in our calorie-rich society. Many of us aren't used to going long periods without eating, like our ancestors likely had to do when food was scarce. Most people don't even go a full 12 hours overnight without eating. They snack late at night and have breakfast early in the morning. And anyway, who wants to starve themselves, even if it means living longer? Do you really want to be hungry all the time, just to eke out a few more years?

Some will say *Yes*, but many of the people I know would say *No way!* Many people graze through their days, essentially eating all day long. It's just what we're used to and what we were taught back in the nineties was good for us. (Remember the graze craze, when so-called health experts recommended we eat lots of small meals during the day instead of a few bigger ones?) But unless you stop eating continuously, you are going to give your body much less time for autophagy.

This is where intermittent fasting can become a more attractive option than calorie restriction. A 2022 study published in the *New England Journal of Medicine*[114] concluded that calorie restriction and calorie restriction with intermittent fasting were equally effective in terms of weight loss and the improvement of many biological markers for health, and another study showed that intermittent fasting *without* calorie restriction was also an effective strategy for weight loss and the achievement of other health benefits.[115] So which would you rather do? To help you decide, let me give you my take on intermittent fasting.

INTERMITTENT FASTING

Intermittent fasting is an easier and, for many, a more comfortable method for promoting autophagy than calorie restriction or prolonged fasting. Studies are showing that even short-term fasts can induce significant levels of autophagy.[116] Fasting has been a part of human civilization for thousands of years, often by necessity during times of food scarcity, but also for spiritual reasons. Fasting is a tradition in many religions, including Christianity, Judaism, and Islam. While some people still practice fasting, for most in the US today, it's just not something we normally do.

But it makes sense. Instead of depriving yourself at every meal, you can fast for 12 to 18 hours per day (including when you are sleeping), reserving your eating time to anywhere between a 12-hour to 6-hour window. For instance, for a 12-hour fasting window and a 12-hour eating window, if you stop eating at 7:00 p.m., you wouldn't eat again until 7:00 a.m.

I recommend starting with a 12-hour window if you've never tried intermittent fasting before. All you have to do is check the time you have your last bite of food for the day, then don't eat again until 12 hours later. Many health experts believe this is the minimum time anyone should fast overnight, to give the body a chance to heal, repair, and engage in cellular cleanup.

Once you can easily fast for 12 hours, you might want to take it further by increasing your fasting time and decreasing your eating window. For instance, you might graduate to a 14-hour fast with a 10-hour eating window between, such as eating only between 8:00 a.m. and 6:00 p.m. If that becomes easy, you can increase your intermittent fast even further. I believe the sweet spot for maximum benefits from intermittent fasting is to fast for 16 hours and eat within an 8-hour window. Let's say you finish eating at 8:15 p.m. In this case, you would not have any food again until 12:15 p.m. the next day. Or, if you like an early breakfast, you could start eating at 7:00 a.m. and finish eating by 3:00 p.m.

Either way, you will probably end up skipping either breakfast or dinner. Although some research suggests that it's better to skip dinner than breakfast, the difference is probably negligible, and what matters more is what works with your schedule. A 2020 study showed no detrimental effects of skipping breakfast in people trying to lose weight,[117] so if you like to make dinner your main meal and you aren't hungry in the morning, that's probably fine. That being said, if eating late compromises your sleep, it's better to finish eating early, and not eat a large amount of food within a few hours of bedtime. In fact, my good friend and the author of *Intermittent Fasting Transformation* (highly recommended!), Cynthia Thurlow, recommends trying to avoid eating three hours prior to sleep.

You also don't need to do this every day. I personally do a 16-hour intermittent fast a couple of times a week, and my preferred way to do this is to stop eating by 8:00 p.m. and then wait to eat until noon the next day. But you do you! Think about which meal is the most important to you. If you are fine with just coffee or tea (with no milk or sugar) in the morning and can make it to noon, you may prefer to skip breakfast and enjoy a larger dinner. However, if you are a morning exerciser and you know you need some fuel, you might be better off moving your eating window earlier. Then, practice your intermittent fasting on days when you feel like you can easily skip dinner, and don't worry about it on days when you know it will be inconvenient, like on a night out or when you are meeting friends for brunch.

If you want to lose weight quickly, reducing caloric intake during your eating window can certainly help. For our purposes—anti-aging and more youthful skin—you don't need to limit how much you eat during your eating window. It's more about *what* you eat on those days. Ideally, you'll eat an autophagy-promoting diet. More on that soon.

What Breaks a Fast?

Fasting means not consuming any calories. That means you can have black coffee or black or green tea in the morning without breaking your fast. However, if you add any cream or sugar, you will be breaking your fast. This will stop autophagy in its tracks. Don't confuse your body with artificial sweeteners, which can disrupt your metabolism and your microbiome. Some people believe that pure fat won't break a fast, such as with a cup of butter coffee (see page 92), and while pure fat is less likely than carbs or protein to disturb autophagy and can help to keep you in ketosis if you are on a ketogenic diet (page 129), it does technically break a fast. If you really can't get by without it though, you will still get some autophagy benefits if you use butter coffee to get you through until lunch.

WHO SHOULDN'T FAST?

While I think that fasting for 12 hours overnight is good for almost everyone, longer fasts are not appropriate for all people. Before beginning an intermittent fasting program like the schedules I've recommend in this chapter, I hope you'll discuss your plan with your physician, especially if you have underlying health issues.

I also don't recommend fasting for anyone in the following situations, unless your physician recommends it:

- **Pregnancy or breastfeeding.** Your baby needs nutrients for proper growth and development. This is not the time to begin a fasting regimen.
- **Underweight.** If you have a body mass index (BMI) of 19 or less, then the likely weight loss from intermittent fasting may be detrimental to your health, especially if you aren't getting in sufficient calories during your eating window.
- **Malnourishment.** If you are malnourished, then fasting may deprive your body of much-needed nutrition. Discuss this with your doctor.
- **Eating disorders.** If you have an eating disorder like anorexia, bulimia, or binge-eating disorder, fasting may not be appropriate for you.
- **Childhood.** Children need a steady supply of incoming nutrients for proper growth and development. The autophagy-promoting diet/intermittent fasting as described in this chapter and in the Jump Start should not be used for children.
- **Gout.** Intermittent fasting could increase your gout symptoms because it can have the effect of uric acid reabsorption. Discuss this with your doctor prior to beginning any type of fast.

- **Other medical conditions.** There are too many medical conditions to name that might potentially be negatively influenced by fasting, so if you have any serious conditions, please discuss fasting with your physician prior to undertaking it.

What about Ketogenic Diets?

A ketogenic diet (or "keto diet") can help you to spend more time in autophagy. However, this type of diet is very high in fat, with the goal of switching your body from burning glucose to burning ketones (which it can do when there are no more available carbs). Burning ketones for fuel is called ketogenesis, and when you're in a ketone-burning state and have detectable ketones in your blood, this is called ketosis. Some studies suggest that ketosis can help people lose excess weight quickly and possibly improve type 2 diabetes, but it can be very hard to sustain, and in people who are fat-sensitive, such as those with heart disease, the ketogenic diet may be hazardous.

Keto diets do have many fans and many success stories, and I believe there are healthy ways to practice keto that can help certain people thrive, such as those recommended by my good friends and holistic health specialists Dr. Anna Cabeca and Dr. Stephanie Estima. They advise their patients and followers on the healthy way to do keto, otherwise known as "clean keto." This is contrasted with "dirty keto" or "lazy keto" which I believe to be more harmful, as it consists of unhealthy sources and types of fat.

Bottom line, a medically supervised "clean" keto diet may be great for certain conditions and people, and does promote autophagy. If you want to try keto and believe it can benefit you, follow the recommendations of holistic health experts. You can learn more about Dr. Cabeca's recommendations at drannacabeca.com and Dr. Estima's approach at hellobetty.club.

EATING TO PROMOTE AUTOPHAGY

While it's true that not eating promotes autophagy, there are also particular foods that just might supercharge your autophagy. Even when you aren't fasting, you may be able to lengthen the time your body spends in autophagy by eating certain foods after you intermittent fast, because while some foods will stop autophagy in its tracks, other foods may promote and prolong it.

To understand how to eat for autophagy, it's useful to know that glucagon supports autophagy, and insulin stops autophagy. This was first demonstrated in 1962, when scientists showed that autophagy in rats was activated by glucagon.[118] Glucagon is basically the opposite of insulin. Eating carbohydrates and protein triggers the release of insulin, while not eating at all triggers the release of glucagon.

Logically then, reducing simple carbs and protein can promote autophagy rather than suppress it with insulin. You can do this by reducing your protein intake to 25 grams or less per day. Since your body doesn't make its own protein, it responds to a protein-specific fast by consuming the proteins stored in cells.[119] In other words, you might say it's eating the protein you already have onboard, starting with the damaged bits. Reducing protein also reduces insulin production,[120] so it can induce autophagy in two ways at the same time.

You don't want to restrict protein all the time. Remember, you need it to support your body's collagen production, which is important for anti-aging. However, on the days that you practice intermittent fasting, you could eat low-protein (and low-carb, since carbs also trigger insulin release) when you are in your eating window. This may lengthen the effects of your intermittent fast by continuing the process of autophagy, even though you are back to eating. However, on days when you aren't intermittent fasting, it's important to eat a moderate to high amount of protein and fiber-rich carbs. You'll see how to do this during the Jump Start, which alternates days of intermittent fasting and an autophagy-promoting diet with days of the Younger for Life Diet, which focuses on nutrients, building collagen, reducing inflammation, and antioxidant action.

Fat has a limited impact on halting autophagy, unlike carbs and protein, which stop it abruptly. This is why I emphasize healthy fats on intermittent-fasting days. That doesn't mean feasting on bacon and butter on every fasting day, though. Instead, remember that omega-3 fatty acids and monounsaturated fatty acids have strong anti-inflammatory effects. Your fasting days are the days to enjoy more avocados, olive oil, grass-fed-sourced butter, nuts and seeds, coconut and MCT oil, and fatty fish like salmon, cod, mackerel, tuna, trout, sardines, anchovies, herring, and krill oil.

And guess what else helps to induce or prolong autophagy? It's those antioxidant foods we've already talked about. It's *almost like* the very best autojuvenating foods also happen to be the very best autophagy-promoting foods! (Coincidence? I don't think so!) Prioritize foods with antioxidants called polyphenols, which have specifically been shown to induce autophagy. These include resveratrol, catechins, quercetin, and curcumin, all of which can be found in bright or dark-colored produce (like plums, cherries, blueberries, blackberries, strawberries, raspberries, artichokes, red onions, and spinach), as well as black beans, hazelnuts, almonds, pecans, coffee, green tea, Earl Grey tea, red wine, and spices like cloves, turmeric, and ginger. You'll notice a lot of the foods on the following list are familiar and make an appearance in other lists. That's because many of the components in healthy foods have multiple positive functions.

AUTOJUVENATING FOOD LIST: AUTOPHAGY-PROMOTING FOODS

FOODS RICH IN HEALTHY FATS (FOR THE BEST FAT PROFILE, CHOOSE WILD-CAUGHT FISH)

Almonds
Anchovies
Beef, grass-fed
Chia seeds
Cod
Eggs from pastured
 chickens
Flaxseed
Hazelnuts

Herring
Krill oil
Mackerel
Pecans
Salmon
Sardines
Trout
Tuna

VEGETABLES

Artichokes
Avocados
Brussels sprouts
Kale

Mushrooms
Red onions
Spinach

FRUIT

Blackberries
Blueberries
Cherries

Plums
Raspberries
Strawberries

SPICES AND CHOCOLATE

Chocolate, dark
Cloves

Ginger
Turmeric

BEVERAGES

Coffee, black
Red wine
Tea, especially green
   and Earl Grey

CHAPTER TEN:

THE YOUNGER FOR LIFE SUPPLEMENT PROTOCOL

My original interest in dietary supplements was as a surgeon looking for ways to help my patients heal better from their surgeries. Many years ago, as I was becoming more interested in holistic and integrative medicine, I started hearing stories from my holistic-practitioner colleagues about how supplements helped them and their patients get off prescription medications. It got me thinking about the use of supplements to potentially help my patients heal better after surgery. Was there a precedent for it? What did the literature say?

Prior to this, when patients asked me before having surgery what they could take to help them heal faster, I would actually instruct them to *stop* their supplements, for fear that one of them might increase the risk of complications such as bleeding. I wasn't taught anything about supplements in my medical training, so I didn't know which ones might help or hurt the healing process. I'm ashamed to say that my ignorance about supplements lasted until I was upward of 10 years into my private-practice career!

Once my interest was piqued, I spent hundreds of hours researching from two very different sources. First, I studied the general surgery, wound healing, and trauma literature to find out which nutritional supplements seemed to help patients heal better. The vast majority of studies about nutritional supplementation and healing were focused on trauma patients who had suffered horrible injuries, putting them in the ICU and making them unable to eat a regular diet. I also found information on chronic wounds (diabetic patients with foot ulcers or paraplegic/quadriplegic patients with pressure sores). These studies focused quite a bit on vitamin C and protein supplementation with arginine and glutamine.

My other source of information was the alternative medicine realm. What supplements seemed to reduce inflammation, support the microbiome (especially important after being on antibiotics), and promote healthy skin? I combined all the information from these two disparate areas into a nutritional supplement protocol that I began to discuss with my patients.

Interestingly, although I didn't conduct any actual studies, I noticed that through the protocol, my complication rate declined. I saw fewer wound infections, and patients seemed to heal faster and more easily than they had before.

Since I created this protocol many years ago, there have been several companies that have created their own pre- and post-surgery supplement systems, and their ingredients are strikingly similar to mine. (Hmmm...) These companies are now advertising their products at surgery conferences around the world. I guess I was on the right track!

Now, if you're not interested in plastic surgery and don't ever plan to have to heal from it, you may be wondering how this story relates to aging. It does, I promise! After starting so many of my surgery patients on the pre- and post-op supplement protocol, I began getting reports that surprised me. My patients repeatedly told me that the supplements I'd recommended were also making them feel and look better in ways totally apart from the results of their surgeries.

They told me how their skin was glowing, their hair was getting thicker and less brittle, their nails were growing stronger, their energy was through the roof, their digestion was improved, and even chronic pain was abating. Of course, as I'm always on a quest to help my patients live better, it got me thinking. Was an age-preventive supplement protocol the next logical step?

When I thought about it, I realized this was a no-brainer. If nutrition or the quality of the food you eat can make a massive impact on the health and quality of your skin, shouldn't nutritional supplements do the same thing?

SHOULDN'T WE EAT OUR VITAMINS?

At this point in my studies, I was already getting more interested in nutrition and thought that we should probably be getting most of our nutrition from food. I still believe that, and it's true that you can't supplement yourself out of a bad diet, but the truth is that nobody eats a perfect diet.

Not to mention, our food is less nutritious than it once was. One study looked at the nutritional data for 43 fruits and vegetables between 1950 to 1999 and discovered a statistically significant reduction in the amount of six key nutrients: protein, calcium, potassium, iron, riboflavin, and ascorbic acid.[121] If our soil was depleted in 1999, just imagine how depleted it is now! Another study looked at 70 diets to determine whether food alone could supply enough nutrients to prevent deficiency. Every single tested diet fell short.[122]

My point is that although the foundation of anti-aging happens through diet, food can only take you so far. If you want a higher-level intervention, consider adding supplements to your diet to target the effects of aging. Taking the right nutritional supplements can be a great way to help your body get the nutrients it needs to keep your skin and body looking younger and can be an integral part of autojuvenation.

This was how I developed my Younger for Life Supplement Protocol. My goal was to include those supplements that have been scientifically shown to improve the skin in various ways, but I also wanted to create a simple regimen that would be easy for someone to follow. I know, as a physician, that there are no miracle cures. But there is real evidence that certain nutrients really do support skin health as well as immune and microbiome health, so that is where my supplement protocol is focused.

By following the Younger for Life Diet, you are doing everything you can to fuel your body with the right foods to slow down aging, but I believe supplements can take you to the next level. So what should you take?

THE YOUNGER FOR LIFE SUPPLEMENT PROTOCOL

Holistic and alternative physicians have recommended all kinds of nutritional supplements for all sorts of reasons. Yet few talk about which supplements to take to prevent and reverse skin aging. Meanwhile, traditional physicians haven't fully embraced the potential benefits that nutritional supplements can have on appearance, not to mention overall health. And I can't tell you how many times I've had muscled-up male fitness trainers in their twenties comment on my TikTok videos, insisting collagen supplements don't work. (These are the same knuckleheads who exclaim that weight loss is as simple as "Calories in, calories out!" and "Milk does a body good!")

This was where I saw a gap in our understanding: professionals in these fields either generally ignore or misunderstand the effect supplements can have on aging, or they focus solely on fixing health problems. But my patients want to look and feel younger. So after years of research, I've come up with a very simple supplement protocol to help your body slow down or even reverse the aging process. My patients confirm that this protocol can help you look younger.

My protocol is very simple. It contains just five products, each carefully chosen to perform a specific task related to anti-aging and keeping your skin looking young. These five supplements are all you need and can be pretty easily purchased, either from my nutritional supplement line or from many other companies. (If you're interested in my products, look to the back of the book at Appendix 3 where you can find information on obtaining them.)

THE RIGHT MULTIVITAMIN

If I had to recommend just one supplement, I would start with a skin-focused multivitamin. Like many of you, I grew up taking a Flintstones multivitamin every day. These little grainy vitamins shaped as Dino, Fred, Barney, and the others tasted like candy. There are better multivitamins out there, especially for adults who want healthier skin. Look for one that contains the following vitamins in particular:

- **Vitamin A.** This fat-soluble vitamin acts as a powerful antioxidant, is critical for eye health, and protects from UV-related photo damage and inflammation.[123]

   (Vitamin A supplements work from the inside out, but also see chapter 13 for how to use topical vitamin A, aka retinoids, for anti-aging from the outside in.)
- **Vitamin C.** You must get this water-soluble antioxidant through food because your body can't make it. Vitamin C is a cofactor that helps to stabilize the triple helical structure of collagen, making it an essential vitamin for healthy collagen production. People who are deficient in vitamin C for long periods of time (historically, think sailors on long voyages) can develop scurvy, which causes skin fragility, petechiae (tiny lesions from bleeding under the skin), bleeding gums, easy bruising, and slow-healing wounds.

Studies show that supplemental vitamin C also acts as a photo-protectant, which means it can help make your skin more resistant to damage from UV radiation and free radicals (but you still need sunscreen!).[124] Another study found that women who had higher intakes of vitamin C had fewer wrinkles and less dry skin.[125] Note that some skin serums also contain vitamin C. It's excellent for topical as well as internal use.

- **Vitamin D.** Vitamin D deficiency is common in the United States.[126] Our bodies manufacture it in response to sun exposure, but you can also take it as a supplement. Vitamin D keeps our bones strong, supports a healthy immune response, and helps prevent skin cancer.[127]

  Unfortunately, our skin synthesizes less vitamin D as we age.[128] That's why it's even more important to supplement with vitamin D as we get older.[129] One study showed a correlation between low levels of vitamin D in the skin and more photodamaged skin, specifically higher levels of erythema (patchy areas of reddening skin), telangiectasias (small red veins), hyperpigmentation (areas of discoloration), and wrinkling.[130]

- **Thiamin (vitamin B₁) and riboflavin (vitamin B₂).** These two essential B vitamins support healthy hair, skin, and nails. They also act as antioxidants, neutralizing damaging free radicals.

- **Niacin (or niacinamide, which is vitamin B₃).** This B vitamin can protect skin against the damage caused by UV radiation.[131] One study showed that it reduced the rate of skin cancers and pre–skin cancers.[132]

- **Pantothenic acid (vitamin B₅).** This B vitamin was shown, in a dermatology study, to reduce facial acne in people who took it as a supplement for 12 weeks.[133]

- **Green tea extract.** This is something I include in my supplement formula that is not present in many other vitamin/mineral supplements, although you can take it separately.

I like it because it contains powerful antioxidants that (as I've shown you previously) can help to induce autophagy. One study found that green tea supplements as well as topical green tea products improved the elasticity of skin.[134] It's also anti-inflammatory and protective against UV-light-induced skin damage and cancer.[135]

POWERFUL ANTIOXIDANT SUPPORT

As I've already talked about, antioxidants can provide defense against free radicals and the harmful oxidative process. Although it's great to get antioxidants from food, it can be difficult to get an optimal level, even with a healthy diet. That's why I value supplementation with antioxidants, ideally including resveratrol, curcumin, and quercetin. In general, the more antioxidants in your diet, in whatever form, the better. You'll fight free-radical damage both on your skin and to your internal organs.

AGE-DEFYING OMEGA-3 FATTY ACIDS

I've already recommended the intake of fatty fish for its omega-3 fatty acids (page 69), but again, it's hard to get enough, and these heart-healthy fats are also excellent for generating younger-looking skin. Studies show that women are less likely to have dry, atrophied skin when they take in more omega-3 fatty acids. Typically, omega-3 fatty acid supplements come in the form of fish-oil capsules. There have been reports of rancidity in fish oil, so be sure you refrigerate it after you open it, mind the expiration date, and buy from a company you trust.

GUT-SUPPORTIVE PROBIOTICS

A daily probiotic is very important to keeping a healthy microbiome and preserving a healthy gut-skin axis. I recommend a probiotic with 5 to 15 billion CFUs. Anything less than that likely has too few organisms to make a positive effect when you consider that the microbiome contains trillions of bacteria. Remember to keep your probiotic in the refrigerator to prevent it from losing potency.

SKIN-FIRMING COLLAGEN

Collagen supplements, which are usually composed of hydrolyzed collagen, are digested into amino acids and peptides in your gut. In order to be used by your body, those amino acids and peptides have to be absorbed by the small intestine and circulated into your bloodstream. One argument against collagen supplements is that it is broken down by stomach acids and is no longer collagen when it is absorbed by the body, if it even is.

However, studies appear to show that this is not necessarily the case. A 2018 study investigated the effects of ingesting hydrolyzed collagen over a four-week period. The results found changes in the levels of hydroxyproline peptides in the bloodstream.[136] A 2017 study found elevated levels of collagen-derived peptides in the skin after supplements of hydrolyzed collagen.[137]

Contrary to what some traditional physicians believe, collagen supplements really can improve skin elasticity and joint comfort, as well as reduce wrinkles and thicken the skin. These claims have been supported by many, many studies. One double-blind placebo-controlled trial showed that ingesting a collagen supplement improved the skin's elasticity in its participants.[138] Another placebo-controlled clinical trial found that oral supplementation with collagen improved the hydration of the skin and increased the density of the collagen in the skin.[139]

A 2014 randomized placebo-controlled study found that oral supplementation for 8 weeks with a hydrolyzed collagen supplement reduced wrinkles around the eyes by 20% and resulted in a 65% higher content of procollagen Type I and an 18% increase in elastin in the skin.[140] Another placebo-controlled clinical trial found that regular collagen supplementation for 6 months significantly improved the appearance of cellulite.[141]

Finally, a 2021 meta-analysis of over 1100 patients found a reduction of wrinkles and improvement in skin elasticity and hydration in subjects after supplementing with hydrolyzed collagen for 90 days.[142] So yes, ingesting collagen can make your skin look and feel younger. As you can see, research vigorously supports this!

There are five types of collagen. Different products contain different kinds and don't always specify.

- Type I collagen is found in bones and skin. It accounts for more than 90% of the organic mass of bone and is also the major component of skin, tendons, and ligaments. This is the type of collagen prevalent in most collagen peptide products, and it's the type that likely has the most impact on skin.
- Type II collagen is the main collagen of cartilage. This type is often taken to support injuries.
- Type III collagen is found mainly in muscles.
- Type IV collagen is less common and aids in the filtration of the kidneys.
- Type V collagen is fiberlike and most commonly found in the placenta.

I advise some caution when choosing a collagen supplement. Some low-quality products contain ground-up, discarded pig, fish, chicken, and cow parts and have a risk of heavy metals. Hydrolyzed collagen supplements (sometimes called collagen peptides) are often basically flavorless and can be mixed into any drink, such as coffee, tea, and even water. Some varieties are also flavored and are good scooped into a smoothie. They also happen to be a good source of added protein.

Do not take collagen during fasting hours, since it contains protein, which can break a fast. It should also not be taken during the Autophagy Diet (see page 186) phase of the Younger for Life Diet, because protein can halt autophagy.

Autojuvenated

Renee is one of my patients. She's 67 years old, and after menopause, she noticed her skin getting thinner. She got bruises from even minor bumps. When she looked in the mirror, she didn't see the face she wanted to see. Instead, all she could see were wrinkles and age spots.

After seeing my YouTube videos on the benefits of ingesting collagen, Renee began taking a hydrolyzed collagen supplement. She put one scoop into her coffee every morning. It dissolved easily and had no taste. After three months of doing this, she was amazed! Her skin looked visibly smoother and tighter, and she began to get compliments from her friends and even her dermatologist. None of them knew she had started taking collagen.

On top of her skin looking smoother, her whole body felt younger. Her joints were no longer sore in the evenings, and her hair was thicker than it had been in years. Taking a simple collagen supplement made a huge impact on her life.

That's it—five types of supplements to autojuvenate your body. Add these supplements in the morning with breakfast and you'll be starting strong and giving your body what it needs to stay energized and youthful. I encourage you to try nutritional supplementation as I've described in this chapter and see how your skin looks and your body feels after using them for a few months.

A final note: There are many great supplement companies out there, but I haven't researched enough of them sufficiently to give you recommendations for brand names. So I encourage you to please check their ingredients and determine what is affordable and available to you. (If you'd like information on my line of supplements, which contains all the nutritional supplementation I've mentioned in this chapter, see Appendix 3.)

PART THREE:

RENEW WITH SKIN CARE FROM THE OUTSIDE IN

CHAPTER ELEVEN:

CLEANSING FOR AGELESS SKIN

I hear from my followers all the time about how they think my skin looks perfect.

"You have glass skin."

"How can I get my skin to look as good as yours?"

"Why is it that you're really old but you don't have wrinkles?"

"What is your ancient Chinese secret to eternal youth?"

Yes, these are real comments made over the years. And no, I'm not Chinese, I'm Korean. Most of my patients don't know this, but I've had issues with my skin all my life. Adult acne, pigmentary blemishes, recurrent perioral dermatitis, and rosacea are just some of the problems I've had to deal with. My skin is also exquisitely sensitive to whatever I put on it. I develop a rash after applying most drugstore-brand skin care products (one reason why I developed my own line—it's one of the few product lines that doesn't irritate my skin).

I want to remind you that I'm not a dermatologist, and during my training, like the majority of young plastic surgeons,

I had virtually no education in skin care. Because I've always been interested in it, though, I decided to learn the basics of skin care as a resident and even gave a talk about it during one of our weekly grand rounds meetings. I remember several of the older plastic surgeons in the audience taking notes.

After completing my general and plastic surgery residency training, I spent one year in Beverly Hills completing an advanced aesthetic plastic surgery fellowship with one of the country's top cosmetic plastic surgeons, Dr. Richard Ellenbogen. It was basically an apprenticeship, so I learned the ins and outs of advanced cosmetic plastic surgery, which I had not yet been taught. In addition to learning these advanced surgical techniques, I finally had the time to devote to learning more about skin care.

I took courses with one of the country's top cosmetic dermatologists, Dr. Zein Obagi, and was later a member of his nationwide faculty for many years. I read textbooks on skin care and cosmetic dermatology. I even took classes on makeup, where I was always the only male in attendance. One Shiseido class I took with my wife unexpectedly required me to apply makeup on her, and her to do the same with me. Believe it or not, I could create a fairly acceptable smoky eye back then!

When I started my private practice in Rochester Hills, Michigan, back in 2004, I made sure to bring in a couple of skin care lines to promote and sell to my patients. The vast majority of cosmetic plastic surgeons sell skin care products in their offices, even though most of them know fairly little about skin care as a whole. Most plastic surgeons sell one of a handful of different brands, including ZO Skin Health, Obagi, SkinMedica, and SkinCeuticals, since the number of skin care companies that reach out to plastic surgeons are few, and these brands are focused on getting visible results, not necessarily just smelling and feeling nice on the skin.

For this reason, I promoted only traditional medical-grade skin care products, all while hiding a secret I was ashamed to admit: many of the products I sold (and my patients loved) caused me to break out in a rash, develop perioral dermatitis, and/or flare up my rosacea when I applied them.

They worked great for my patients, because they had active ingredients like retinol, kojic acid, and vitamin C, but I could never use them or any products I knew of. Instead, I stuck with simple cleansers and maybe a light moisturizer for super-sensitive skin. This went on for *years*.

Then, as I began to discover how food and nutritional supplements can improve the skin, I began to look into natural and organic skin care products. To my surprise, unlike drugstore brands or medical-grade skin care products, these did not cause me to break out or erupt in itchy hives. Instead, they left my skin feeling hydrated, comfortable, and *normal*. I realized that the other products I'd been using and selling contained unnecessary chemical additives (not surprising if you look at the dozens of unpronounceable ingredients on their labels), and these were what caused my skin to react poorly. It wasn't necessarily due to my skin being extra sensitive but to the products having ingredients that were extra irritating or reaction-inducing. This was a huge realization for me.

I decided to take a deep dive into skin care ingredients and their potential for adverse reactions. I learned that although the European Union has banned over 1000 ingredients from personal-care products for safety reasons, the US Food and Drug Administration has banned only 11. There are ingredients in our skin care that, when used in large amounts, may increase the risk of cancer, mimic hormones in your body, and cause irritation and allergic reactions. Some of these chemicals are present in the medical-grade products I was selling and promoting in my office!

Armed with this new information, I was excited to start educating my patients about organic skin care. But very quickly I came to another realization. Although these natural and organic products were gentle and great for moisturizing the skin, they weren't as effective in other ways. They didn't actually *do* much to intervene in the aging process.

So I started my own skin care line, YOUN Beauty, to combine natural and organic ingredients with scientifically supported medical-grade components, like retinol, vitamin C, kojic acid, and hyaluronic acid. Of course, you don't have to use my products (Appendix 2 contains a list of other skin care companies I recommend, and Appendix 3 contains information on my own product line, if you're interested), to reverse aging.

Whatever products you decide to use, I do suggest that you treat your skin from the outside in with three basic steps to younger-looking skin, which I'll cover in this and the next two chapters. Healthy skin is attractive skin, so the goal of any skin care regimen should be a health-first approach. By supporting the health of your skin, you will naturally reduce things that may bother you about its appearance, such as wrinkles, age spots, rough texture, and crepey skin.

HOW TO CARE FOR YOUR LARGEST ORGAN

Skin is the largest organ of the body, covering almost two square meters on average. The skin of our entire body ages, but it does so at much different speeds, depending on how much it's exposed to the elements. Compared to the skin on your face, which you've probably at some point treated with cleansers, moisturizers, and sunscreen, the skin of your hands likely looks older.

To take better care of the skin on your face, I recommend that you consider skin care in three different steps, all of which are important: cleansing, protecting, and treating. Let's begin with cleansing.

CLEANSING YOUR SKIN

Most people should cleanse their skin twice a day, but if you're only going to do it once, do it at night. It's very important to wash off pore-clogging makeup and the day's pollution, oil, and dirt that has built up on your skin. The cleanser you choose should leave your face feeling clean and comfortable. It shouldn't make your face red or dry or leave it stinging. And, God forbid, do not use regular bar soap. This often contains the surfactant sodium lauryl sulfate, which can leave a dry film on your skin. Here is some more guidance, based on what kind of skin you have.

- **For the driest, most sensitive skin**, use surfactant-free cleansers that don't require rinsing.
- **For slightly sensitive, slightly dry skin**, consider trying a creamy cleanser with botanicals like green tea, which will leave your skin feeling moisturized.
- **For normal to oily skin**, try foaming cleansers. These are more aggressive but very effective at cleaning the skin and removing makeup. They can also be drying and can irritate sensitive skin.
- **For people with acne**, try an oil-based cleanser. Although not for everyone, some people with oily, acne-prone skin find oil-based cleansers work the best for them. These can leave the skin feeling clean but not oily. Although it sounds counterintuitive, oil-based cleansers are very effective at cleaning oily skin because, as the saying goes, like dissolves like. In other words, oil cleanses oil. These cleansers can efficiently get into oil-clogged pores and clean them out.

You don't need fancy tools to wash your face. With your hands, apply your cleanser with circular motions. Rinse with warm water, then pat dry with a clean towel. Consider doing a double cleanse if you find that cleansing your skin once doesn't remove all the day's makeup and oil. Some people find that an oil-based cleanser is a great choice for double cleansing: start the cleansing process by removing dirt, oil, and makeup with an oil cleanser. Follow it up with a more traditional gentle cleanser to reveal the cleanest, softest skin.

Do You Need a Toner?

What do you do with clean skin? The next step is to prepare it for taking treatments. You may be accustomed to using a toner for this purpose, but is it really necessary? In the past, toners were traditionally used to mop up the oily residue some cleansers would leave behind. Toners were astringents that would make the skin feel cool and clean but could also dehydrate and strip it of its essential oils.

Now we understand even more about how the skin works. We've recently discovered that, like the gut, the skin contains its own complex microbiome. Yes, there are trillions of bacteria living on the surface of your skin, and these healthy bacterial colonies likely play a huge role in your skin health. Stripping the skin of its microbiome using alcohol-based toners or aggressive cleansers can actually make your skin less healthy and more aged-looking afterward.

Just because something feels good in the moment doesn't mean it's good for your skin. As long as your cleanser doesn't leave behind significant residue, then you could probably skip the toner. However, if your skin doesn't feel completely clean after cleansing, you might like how a toner works for you. You may have to experiment.

Toner is probably best for people with oilier skin. Look for a version that doesn't contain alcohol. If you have oily skin, apply toner twice a day or as your skin tolerates. For normal or combination skin, applying once a day is probably fine. Avoid using toner if you have very dry skin, eczema, or rosacea.

Cleanliness may be next to godliness, as they say, but the next step is equally important for preserving and maintaining youthful, firm, and resilient skin: protection.

CHAPTER TWELVE:

PROTECTION FOR AGELESS SKIN

Cleaning and preparing skin are matters of maintenance and preparation for the next anti-aging action: protecting your skin from an onslaught of free radicals and environmental assaults. This is critical for autojuvenation, and anyone concerned with preserving or reclaiming youthful-looking skin should have a protection plan. Protection is three-pronged. For maximum protective effect, you need a good antioxidant serum, sun protection, and moisturizer. Let's look at these individually.

ANTIOXIDANT ACTION FOR ANTI-AGING

I've already talked about eating antioxidant-rich foods and taking antioxidant supplements, but antioxidants are also anti-aging powerhouses when applied topically. As you surely know by now, antioxidants prevent damage from free radicals by neutralizing them, and free radicals are one of the primary perpetrators of skin aging. When you take antioxidant supplements and use antioxidants topically, that helps prevent wrinkles and protects against skin cancer. Think of it as a one-two punch!

Apply an antioxidant cream or serum in the morning so it can protect your skin during the day. It's less important to apply at night when your skin is not being exposed to environmental onslaughts. There are many types of antioxidants that can be applied topically, including vitamin C, vitamin E, green or black tea, coenzyme Q10, and pomegranate.

Vitamin C is the most popular and easiest-to-find ingredient in antioxidant skin serums. Most major skin care manufacturers have a vitamin C cream or serum. Look for one containing at least 10% L-ascorbic acid. Because vitamin C is unstable when exposed to light, it must be packed correctly in a dark container or one that doesn't allow light to penetrate it. If the product is brown, it has oxidized and may not be effective.

But a fresh, high-quality vitamin C serum *is* effective, especially when combined with vitamin E. A study in the *Journal of the American Academy of Dermatology* found that the two can be synergistic.[143]

Although antioxidants are known mainly for their skin-protective benefits, some of them also have potent anti-aging effects. Vitamin C helps reduce unwanted age spots and helps with fine lines by stimulating the production of collagen. However, I like to focus mainly on the protective effects of antioxidants, since fighting free radicals and reducing inflammation are their primary benefits. Even teenagers can benefit from antioxidant serums to help preserve their skin, so I recommend everyone to start using an antioxidant serum as early as possible.

EVERYTHING YOU NEED TO KNOW ABOUT SUNSCREEN

If you want to prevent premature skin aging, it's mandatory to use sunscreen, especially on your face, neck, and upper chest, but ideally also on your hands and arms if they are exposed, every single day. Believe it or not, as much as 60% of the sun's damaging radiation penetrates the clouds, even on really cloudy days.

A study of identical twins from Case Western Reserve University showed that increased sun exposure and lack of sunscreen use led to an older appearance in one twin as compared to the other, and that this difference accelerated with age.[144]

It can take 10 years or more for the aging effects of the sun to surface on the skin. This often shows up as dark spots (age spots). I have many patients in their forties and fifties who complain of age spots but tell me they never get sun exposure and always wear sunblock. I believe some of these spots are the result of sun damage from years or even decades earlier.

The damaging radiation from sun exposure consists of UVA and UVB rays. UVA rays cause aging, discoloration, and wrinkles and can progress to skin cancer. UVA rays have a longer wavelength than UVB rays, so they can penetrate deeper into the skin at a molecular and cellular level. UVA rays are what cause melanoma: they react with skin cells to produce free radicals, which can destroy DNA, sicken healthy cells, and damage collagen and elastin in the skin to cause premature aging and sun damage, as well as contributing to skin cancer. This damage can be even worse when combined with environmental free radicals, such as from pollution.

UVB rays cause suntans and sunburns. Long-term exposure to UVB rays is linked to basal cell and squamous cell carcinoma. These rays are most intense between 10:00 a.m. and 3:00 p.m., so try to limit your sun exposure during this time. Also keep in mind that the sun's rays are more intense near the equator and at higher altitudes.

And did you know that being surrounded by snow can reflect the sun's rays and make it much easier to get burned? I once went skiing in the Alps while in college. Having almost never been sunburned due to the melanin in my skin, I just slapped some sunscreen on my cheeks and hit the slopes. I didn't realize that the high altitude combined with the reflection of the sun's rays off the snow would cause me to have the worst sunburn of my life.

And even worse, my entire face was burned to a crisp except for two handprints on my cheeks! That was a huge lesson I learned, and months of embarrassment afterwards as everyone called me Panda Youn. So please wear sunscreen while skiing and enjoying other outdoor winter activities.

The level of protection in a sunscreen is designated by its SPF, which stands for sun protection factor. However, what many people don't realize is that the SPF number only refers to protection against UVB rays, not UVA rays. Therefore, while a high-SPF sunscreen may prevent tanning and burning, it may not protect you against premature aging, age spots, wrinkles, and potentially deadly melanoma.

To get full protection from both types of rays, look for sunscreens labeled *broad spectrum*. The FDA has implemented a rule that a warning now be placed on sunscreens that lack adequate UVA protection. I also recommend an SPF of at least 30, which will absorb 97% of the sun's rays. This is also the recommendation from the American Academy of Dermatology.

The last thing to know about sunscreen is that SPF numbers are based on the assumption that you are applying a good amount of sunblock on your skin. Most people only apply 25% of the recommended amount. Try to apply one full ounce (about the amount that would fit into those medicine cups that come on top of bottles of cough syrup and cold medicine) to cover your entire body, as recommended by the American Academy of Dermatology. At least a teaspoon of this should go on your face, ears, and neck. If you have short hair, apply to your scalp as well.

If using a chemical sunscreen, apply it 15 to 30 minutes prior to heading outdoors, to allow it to be absorbed by your skin. Reapply every two hours until you go indoors, more often if you're sweating or in the water. And don't forget to apply it to your lips! Lips don't tan, so they are completely exposed to the effects of the sun. Use a lip balm with SPF 30 or higher, then apply your lip gloss or lipstick. Remember, no sunscreen is truly waterproof, even if the label says so. You need to reapply it after being in the water.

SUNSCREEN VS. SUNBLOCK

People often use the terms *sunscreen* and *sunblock* interchangeably, but they aren't actually the same thing. Sunscreen (sometimes known as chemical sunscreen) contains active chemicals that absorb the sun's ultraviolet rays. Sunscreens are absorbed into your skin, and studies show that they are also absorbed into the bloodstream. Chemical sunscreens are thinner, and most people prefer using them on their faces so they don't have that ghostly white hue and heavy, sticky feeling caused by the ingredients in some sunblocks.

While sunscreen can certainly work to protect your skin from the sun's damaging effects, I don't recommend using chemical sunscreen on your entire body because of the chemical exposure. For instance, oxybenzone is a very popular ingredient in chemical sunscreens, but it is a suspected endocrine or hormone disrupter. According to the EWG, 600 sunscreens sold in the US contain oxybenzone.[145] In 2008, top scientists from the Centers for Disease Control (CDC) found oxybenzone in 97% of the general American population's urine, so we know that it gets absorbed through the skin into the bloodstream.[146] On top of that, studies show that 1 in 4 to 5 people have allergic reactions to oxybenzone.[147]

Studies also show that oxybenzone, when exposed to sunlight, creates free radicals,[148] so while it may be protecting you from free-radical formation caused by sun exposure, it could also be creating reactive free radicals through a chemical reaction with the sun. What? Where's the logic in that? Oxybenzone also seems to have weak estrogenic[149] and antiandrogenic effects.[150] Many holistic health professionals consider it a hormone disrupter that mimics the shape of hormones in the body. One study found lower birth rates related to oxybenzone.[151]

Octinoxate, another common sunscreen ingredient, is also believed to disrupt hormones in a similar fashion as oxybenzone. Two studies have shown possible alterations in thyroid function in rats due to octinoxate.[152] Yes, I know it pertains to rats, but still.

On top of these potentially harmful qualities, oxybenzone and octinoxate sunscreens may also be environmentally destructive. They're banned on many beaches near coral reefs, including in Hawaii. It's believed that these sunscreens result in bleaching of coral reefs.[153]

Mexoryl SX and stabilized avobenzone are chemical sunscreens that provide good UVA protection, and unlike oxybenzone, they do not appear to create hormone disruption and have very limited skin penetration. If you do choose to use chemical sunscreen, look for these ingredients instead of oxybenzone or octinoxate.

But what about sunblock? Sunblock contains either titanium dioxide or zinc oxide. These minerals will physically prevent the sun's rays from reaching your skin by blocking them. Instead of a chemical reaction, they act like a shield. However, these very minerals are what create the white residue people often dislike. Fortunately, many companies are now coming out with newer formulas that use a micronized physical sunblock, which has less white residue. It's more expensive, but the better look and feel along with the absence of harmful chemicals may be worth it to you.

There is one concern about sunblock, however. Physical sunblock that is clear, without the white residue, may contain nanoparticles. There are some early signs that zinc nanoparticles may get absorbed by the bloodstream. To be totally safe, you may want to stick with physical sunblock that comes out of the tube white.

Overall, I recommend using physical sunblock for your body, to play it safe. Use Mexoryl SX or avobenzone for your face, or a mineral sunblock if your skin is light enough not to be adversely cosmetically affected by any residue from the cream.

MAKING THE MOST OF MOISTURIZER

Moisturizers are great to prevent water loss from the skin and keep it hydrated, smoother, and less wrinkled, but a moisturizer won't actually make your skin cells younger. It only makes your skin *look* younger, smoother, and healthier (but that's something!). As we age, our skin loses some of its hydration and plumpness. Moisturizer won't reverse aging, but it's essential for preserving what you have.

Ideally look for a daily moisturizer with active ingredients like antioxidants (vitamins C, E, and/or green tea). Even if you already use a serum with these ingredients, it doesn't hurt to have an extra boost in your moisturizer. Apply these moisturizers in the morning to reap the protective effects of the antioxidants.

It's best to apply moisturizers to the skin within two minutes after you shower and when your skin is still damp, in order to lock the moisture into your skin. At night, a more powerful moisturizer can help to restore moisture to your skin and rejuvenate it. You can get a separate moisturizing night cream or use a combination cream—one that also contains active anti-aging ingredients like retinol, peptides, and growth factors.

By the way, if you have oily skin, then you might not need a moisturizer.

Good moisturizers are most important in the winter, when the humidity drops in cooler climates and the skin gets drier. If you have very cold winters, like we do in Michigan, make sure to also use a humidifier, especially while you sleep. This can really help your skin stay hydrated. If your furnace has a humidifier attached to it, make sure it is set to *Winter*, otherwise it won't work.

Dr. Youn's Pro Tip

Here's a great tip for keeping your skin moisturized. Frequently refresh your skin during the day with spritzes or sprays of spring water. This can help keep your skin moisturized, reducing the wrinkles that appear when the skin dries. Many skin care companies have small spray bottles that you can put in a bag or purse and use to keep your facial skin hydrated.

After cleansing and protection, the next step is treating, and this is where skin anti-aging gets really proactive. Let's autojuvenate your skin!

CHAPTER THIRTEEN:

TREATMENTS FOR AGELESS SKIN

Treating your skin is the final of the three steps for tackling aging from the outside in, and it comes in two parts. First, exfoliating the skin to get rid of dead skin cells so the skin not only looks clearer and younger but can better absorb potent anti-aging treatments; and second, the treatments themselves, of which there are many choices. Let's start with looking at exfoliation.

MAKE WAY FOR YOUNGER SKIN
WITH EXFOLIATION

A large portion of the upper layers of skin is composed of dead or dying skin cells. These dead cells can cause a flat or dull look to the skin. Exfoliating will help remove the upper layers of dead skin and speed up cellular turnover, which starts to slow down in your twenties and thirties. Basically, exfoliation causes the skin cells to send signals to produce new ones.

There are two main ways to exfoliate: physical and chemical. Physical exfoliation works via tiny particles you scrub over your skin that gently abrade its upper layers. This is the most budget-minded way to exfoliate, but be very choosy if you decide on this route. Make sure the scrub feels soft and sandy, without sharp particles that can cause damage.

You don't need to scrub hard. Let the texture of the exfoliation scrub do the work for you. Some doctors believe that aggressive physical exfoliation can create microtears in the skin which long-term can result in increased inflammation and unnecessary skin trauma. If you have sensitive skin, then a physical exfoliator may not be your best option. Instead, you may want to go with chemical exfoliation.

Chemical exfoliation uses alpha or beta hydroxy acids to exfoliate. You could try an over-the-counter (OTC) alpha hydroxy acid peel to see if you like this method. Light chemical exfoliation is the preferred method for people with sensitive skin, issues with breakouts, and mature skin, which can be thinner and more delicate.

If you have sensitive skin, I recommend you exfoliate no more than once per week. Normal skin can exfoliate two or three times per week. Slow down with exfoliation if your skin gets red and irritated.

Typically, it's best to exfoliate after you cleanse, so you treat a clean, fresh face. Nighttime is usually the best time for exfoliation since it allows for better penetration of your night creams and active treatments. And just in case you were considering it, do not exfoliate your eyelids! They are typically too sensitive to tolerate the abrasion.

Dr. Youn's Pro Tip

There is a trend right now for less aggressive treatments in skin care. If you're not seeing the results you're looking for or your skin is breaking out, then try eliminating some of the products you're using, especially exfoliators and toners. Go back to the basics—cleansing, sunscreen, moisturizer—and see if your skin improves. Quite often, less is more when it comes to skin care.

After exfoliation is the best time to apply anti-aging treatments, but even on days when you don't exfoliate, anti-aging treatments are best applied at night, so they can penetrate the skin and work their magic while you sleep. Let's look at the best treatments—the ones that actually make a difference in the firmness, elasticity, clarity, and tone of skin. These are absolute keys to the process of autojuvenation.

THE GOLD STANDARD: RETINOIDS

Retinoids are a type of vitamin A that chemically exfoliate skin and stimulate production of new collagen fibers to replace the old, damaged, irregular, and aged collagen fibers. Retinoids can reverse the thinning of the collagen in the dermis that occurs with aging and also have an anti-inflammatory effect. Some studies show that prescription-strength tretinoin or Retin-A can even reverse early pre–skin cancers such as actinic keratosis.[154] (If you've been diagnosed with actinic keratosis, talk with your dermatologist about what is the best treatment for you, as they will likely prescribe you something even stronger than tretinoin.)

The best time to apply retinoids is at bedtime. Allow six to eight hours for best effect. Know, however, that if you use retinoids, especially prescription-strength ones like tretinoin, your skin will be more sensitive to sunlight, so it's extra important to stay out of the sun or wear a strong sunblock.

If you opt for a prescription for Retin-A (tretinoin), you may see visible results within four to six weeks, although try to give it at least three to four months to achieve really noticeable changes. A less irritating form of tretinoin is Retin-A Micro. More moisturizing forms are Renova and Refissa.

To apply these prescription-strength retinoids, you need only a pea-size amount on your facial skin every other night to every night, or as directed by your doctor. Only apply it at night because it is deactivated by light. People often develop irritation, dryness, flaking, or a temporary dermatitis from these products because they are so powerful. You can even develop pimples as your skin purges or acclimates to the treatment. If you develop these symptoms, consult your doctor, who may recommend that you decrease the frequency of use until your skin acclimates.

Many people start by applying it every other night until their skin adjusts and then increase to nightly use. But the more often you use these products, the quicker you will see changes. So work with your doctor to develop a plan that works for you and your skin. Keep in mind, these products are highly effective, as long as you stick with them.

That being said, they're not for everyone, as they may require attention to dosage and an adjustment period. The first time I used tretinoin was as a medical student. My then-girlfriend, Amy, and I began developing adult acne. I've never had a problem with acne before, and this was the first time I felt like I was truly breaking out. So we saw our family doctor at the time, who prescribed each of us 0.1% tretinoin and instructed us to apply a pea-size amount every night.

Knowing what I know now, I'm sure that the acne was due to a combination of stress, poor diet (my choice in late-night hospital cafeteria food was somehow always deep-fried), and lack of sleep, but I don't blame her for not knowing this. She told me that it worked great for her acne and prescribed us her regimen.

But here was the problem: my doctor had very thick, oily skin; Amy and I do not. Thick, oily skin tends to handle the acute effects of tretinoin much better. Naïve of this fact, I began applying what is essentially a maximum prescription-strength retinoid onto my skin, and ten days later my face felt and looked like it was on *fire*. It was bright red, dry, cracked, and shedding white flaky skin all over the place. I was a mess. The next day I saw Amy, and she was a mess, too! We immediately discontinued the tretinoin and started ourselves on a light topical steroid and aggressive moisturizing.

There are some people who simply can't tolerate tretinoin at all, but many of the people who discontinue its use have just given up too quickly. However, if you have thin or sensitive skin, like Amy and I do, I usually recommend starting with a less concentrated over-the-counter retinol cream first. Once you adjust to this, you can work your way up to prescription-strength tretinoin if you want to. Most people achieve nice changes with retinol and never have to progress to a prescription.

Note that if you have very dark skin, you should consult with your plastic surgeon or dermatologist before using retinoids to see if these will work for you.

RETINOL: OTC RETINOIDS

The main advantage to retinol is that it's available without a prescription and is less irritating than full-strength tretinoin. That makes retinol a great place to start with anti-aging treatments. Studies show that even though retinol isn't as strong as prescription-strength tretinoin, it can have a similar anti-aging effect.[155] Retinol is converted in the skin to retinoic acid (tretinoin is one kind of retinoic acid) in very small amounts after it is applied to the skin.

Apply retinol cream every night or every other night to start. Do not bother applying it in the morning since, like retinoids, sunlight can deactivate it. If you pick only one anti-aging cream to use, I recommend it be a retinol moisturizer with natural and organic ingredients. Unless you have extremely sensitive skin, this is a great cornerstone for any anti-aging skin care regimen.

Important: do not use retinoids if you are pregnant or breastfeeding. Although the data is scarce, it's possible that both tretinoin and retinol in topical products can be absorbed into the bloodstream and theoretically increase the concentration of vitamin A in the body to levels that may be harmful to the fetus. Instead, use pregnancy-safe products like moisturizers containing hyaluronic acid, green tea, and/or glycolic acid.

GROWTH FACTORS

Growth factors are chemical messengers that signal skin cells to increase the production of new collagen. One type of growth factor in skin creams is Transforming Growth Factor Beta (TGF-ß). If you cannot tolerate retinoids, lotions with TGF-ß cause much less irritation and may be a reasonable alternative. However, the problem with these growth-factor creams is that they are very expensive.

PEPTIDES

If you don't want to shell out the cash for growth factors but your skin is too sensitive for retinol or even AHAs, you might consider a peptide-based anti-aging cream. Typically, peptides are more easily tolerated, but the results aren't as impressive as with retinoids. Peptides signal skin cells to increase collagen production. Remember, the collagen in our skin degrades as we age, so replenishing collagen is a key goal in reversing the aging of our skin.

Theoretically, peptides can also reduce inflammation and reverse damage from environmental toxins. Palmitoyl pentapeptide, oligopeptide, copper peptide, Matrixyl, and Dermaxyl are all peptides commonly used in skin creams. Other types, namely GABA (gamma-aminobutyric acid) and DMAE (dimethylaminoethanol), can theoretically relax smooth muscles (the kind of muscle that is under the skin, among other places), causing the skin to look less wrinkled. This effect is temporary, lasting only a few hours, but it's an inexpensive way to get an immediate tightening and smoothing of the skin.

If I had to rank all these treatments in terms of effectiveness, I would put retinoids at number one, followed by growth factors, then peptides. If you are a skin care enthusiast and want to go the extra mile, combine retinoids with either growth factors or peptides for optimal results and an even better autojuvenating effect.

Dr. Youn's Pro Tip

Bakuchiol is one of the newest and most talked about anti-aging ingredients in skin care products today. Billed as a plant-based, natural, and gentler alternative to retinoids, bakuchiol is an extract derived from the babchi plant. It's been used for hundreds if not thousands of years in Ayurvedic and Chinese medicine due to its skin-soothing and anti-inflammatory properties.

Although retinoids remain the gold standard for anti-aging skin care ingredients, a 2019 study found no difference between retinol and bakuchiol in the treatment of wrinkles and hyperpigmentation.[156] Although both products had similar results, those who used retinol experienced more skin dryness and discomfort. Like retinol, the effects of bakuchiol on wrinkles appear to stem from its impact on increasing the production of collagen in the skin. But unlike retinol, bakuchiol doesn't appear to sensitize your skin to the sun.

So do I believe bakuchiol is a reasonable alternative to retinoids for people who want youthful, healthy skin? At this time, if I had to choose one, I'd still go with a retinoid due to its long-proven history of anti-aging effects. However, if your skin doesn't tolerate retinoids, then bakuchiol is a very interesting alternative and definitely worth a try.

Or better yet, why not use both?

BRIGHTENING TREATMENTS

One of the most common complaints I hear is about age spots, sun spots, and liver spots. These are all names for the same thing: unwanted pigmentary blemishes caused by excess sun exposure. The UV radiation from the sun damages the pigment-producing cells in the skin (also known as melanocytes), causing them to overproduce melanin and deposit it in clumps in sun-exposed areas. In general, these spots will not go away on their own. They can sit there for years, accumulating and increasing in numbers, getting more and more annoying. The only way to get rid of sun spots is to actively remove them.

The main function of brightening creams (also called lightening creams) is to reduce the appearance of unwanted pigmentation, such as sun spots. There are many kinds out there, with many different active ingredients. Let's get into it.

The most aggressive and effective ingredient in topical brightening creams is hydroquinone (HQ). Unfortunately, because of its potentially toxic effects, I don't give HQ my wholehearted holistic recommendation. However, it's the most potent brightening ingredient there is and is used in some of the most prominent skin-lightening creams, so you should know about it.

Hydroquinone inhibits tyrosinase, a key enzyme essential in producing melanin (the pigment in the skin). A standard strength formula is 2%, and until the CARES Act in September 2020 banned its over-the-counter sale, it was available through many skin care brands. Currently, the only obtainable topical hydroquinone in the United States is 4% strength, which must be obtained via a prescription or through a doctor's office. Hydroquinone is also banned in many European countries.

The best way to use hydroquinone is to combine it with retinoids and AHA. However, a risk is ochronosis, which is a very rare skin darkening that can happen in people using hydroquinone. This complication occurs mainly in people with darker skin.

There is also a concern that hydroquinone may be carcinogenic. It has been shown to cause cancer in laboratory rats when they are exposed to very high concentrations, but this correlation hasn't been proven in humans.[157] Hydroquinone may also have significant rebound hyperpigmentation when stopped cold turkey.

Because of the above concerns, I recommend all my patients who use hydroquinone to stop using it within six months and convert to a less potentially toxic option (like kojic acid). Do not use hydroquinone-containing products if you are pregnant or nursing. Better yet, avoid using these at all. Instead, consider combining a skin brightener that doesn't contain HQ with intense pulsed light (IPL) treatments (see page 275), for a quick result without the possible toxicities.

I personally prefer kojic acid to HQ. It's a skin brightener that also works by inhibiting tyrosinase. It's slightly more irritating than hydroquinone, but it's less expensive and doesn't come with the risk of ochronosis. It's also obtainable without a prescription. But full disclosure: a recent study in the *Indian Journal of Dermatology* found that hydroquinone was more effective than kojic acid for the treatment of melasma (which is a skin condition that creates dark patches).[158]

Because kojic acid is less effective than hydroquinone, it's best combined with an exfoliating agent for better penetration, such as an AHA or retinoid. Or better yet combine it with IPL (see page 275).

Another alternative skin brightener is niacinamide. This is a common ingredient in drugstore-brand skin brighteners. Niacinamide is a form of vitamin B that is shown to help with wrinkles and pigmentation. It doesn't appear to be as effective as hydroquinone or kojic acid but is a good budget-minded alternative that still has an effect. Niacinamide also has no known toxicities associated with its use, unlike hydroquinone.

Skin Care Ingredients to Avoid

What's irritating your skin when you put on your favorite skin care product? It might be the added or unnecessary chemicals. It's amazing to me that we allow some of these things into our skin care—especially potential carcinogens and hormone disrupters—but we do, so here's what you need to know about the ingredients I recommend that you avoid:

- **Parabens (isopropyl, butyl, isobutyl).** Parabens are used as preservatives in many skin care products. They are known xenoestrogens, mimicking estrogens in the body. They can have hormone-altering effects and may even increase the risk of breast cancer.[159] In men, urinary paraben concentrations were significantly associated with an increase in the percentage of sperm with abnormal morphology, a decrease in sperm motility, and decrease in testosterone levels.[160] One study found that most people have some level of parabens in their body.[161] (Note: the type of paraben is important. Methylparabens have been found to overall be safe without evidence of toxicity or hormone-mimicking effects.)
- **Ethanolamines (DEA, TEA, MEA).** Ethanolamines are used to make skin care products feel creamy, but studies show exposure to high levels of these chemicals is linked to liver cancer and precancerous changes in the skin and thyroid.[162] Both the EU and Canada classify DEA as toxic, but it's still used regularly in the US. Ethanolamines can also react with certain preservatives like N-nitrosating agents to form nitrosamines, which are considered potent carcinogens.[163]
- **BHA and BHT.** These chemicals are used to extend shelf life. They are likely human carcinogens[164] and hormone disrupters.

171

- **Phthalates (DBP, DEHP, DEP, and others).** Phthalates are a class of plasticizing chemicals used to make products more pliable or to make fragrances stick to skin. Phthalates disrupt the endocrine system and may cause birth defects[165] and cancer.[166]
- **Polyethylene glycol (PEG compounds).** PEGs are widely used in cosmetics as thickeners and moisture-carriers. They may be contaminated with ethylene oxide and 1,4-dioxane, which are both carcinogens.
- **Formaldehyde releasers (quaternium, diazolidinyl urea, DMD hydantoin).** Formaldehyde is known to cause DNA damage and cancer.[167] In addition to its use in skin care products, formaldehyde is also present in building materials, pressed-wood furniture, glues, and vehicle exhaust. A study of funeral-home workers found that those who did more embalming (and therefore were exposed to more formaldehyde) had an increased risk of dying from myeloid leukemia.[168] We are all exposed to formaldehyde in some form, so I advise against worsening the situation by putting it on your face.

CHAPTER FOURTEEN:

SKIN CARE ROUTINES FOR YOUR AGELESS LIFE

I have friends in the holistic space who are gut-health experts, and whenever someone asks about what they can do to improve their acne or wrinkles, they recommend things that will improve their gut health and overall health, like taking a probiotic, drinking bone broth, or eating the rainbow of fruits and vegetables. They usually don't make any mention of skin care products at all.

On the flip side, I have dermatological colleagues and friends who answer this same question with solely the skin care products you should use. If you have wrinkles, apply a retinoid. If you have sun spots, use kojic acid and niacinamide. If you have rosacea, use a prescription cream like Soolantra. They usually don't mention diet or gut health.

Well, I'm not a gut-health expert. My holistic-medicine colleagues know a lot more about that than I do, and I've learned what I know about the effects of our gut health on our skin and body from them. And I fully admit my dermatology colleagues know more about skin conditions than I do. I've learned almost everything I know about skin care from those colleagues, to whom I owe a debt of gratitude.

However, combined, these two very different groups of health experts have also taught me what's missing: a truly integrative approach is the absolute best way to give yourself healthy, younger-looking skin.

I've been saying this from the beginning of this book. It's not just about what you can do from the inside out, or from the outside in. The most powerful approach is to tackle aging on both fronts, a key principle in autojuvenation. It's not enough to eat the right foods or take the best nutritional supplements. This won't make your lines go away or significantly lighten those dark spots. Conversely, applying the right creams without addressing the underlying root cause of why a person has skin conditions or premature aging is only improving the symptoms and not fully treating what's going on underneath the surface. Once you stop applying those creams, the skin condition that you are treating—whether it's acne or premature aging or whatever—will come back.

It's the same when you look at so many chronic-health issues that we are dealing with today. Someone may be prescribed a medication to treat their hypertension, but if they don't get to the root cause of why they are dealing with hypertension (including diet, activity level, stress, sleep, excess weight, and genetics), then once they stop the medication, the hypertension will come back. Of course, medications may be necessary to bring down the high blood pressure more quickly than lifestyle changes will in order to prevent the person from having a stroke, but taking a medication certainly doesn't preclude making lifestyle changes to support healing and to correct the root causes of any health issue.

I want to remind you of all of this before I get into taking what you've learned in this section about skin care and putting it into a routine. I love these simple anti-aging protocols, but they are only about what you're doing on the surface. Pick your favorites and practice them religiously, and you will see a difference. But don't forget about the other parts of this book: while you do these skin care routines, remain mindful and vigilant about your diet, your sleep, your stress, and all the other things I'll be covering soon.

THE TWO MINUTES, FIVE YEARS YOUNGER SKIN CARE ROUTINE

This is my most popular routine. It's a very simple skin care routine that I recommend to my patients. It takes just two minutes every morning and evening, but it can keep your skin looking healthy and even make you look five years younger within a couple of months. In fact, my team conducted a simple study of female patients who tried this routine. We took photos of them prior to starting the routine (you can see these on my website at younbeauty.com) and then again two months later. When we polled people, asking them how much younger they felt they looked, the majority of respondents answered, "Five years younger or more!" And so this routine got its name. By performing these simple steps, in as little as six to eight weeks, your skin could look five years younger. Here is that exact routine. (For a list of skin care brands that I recommend, please check Appendix 2.)

DAILY
Morning, Step 1: Cleanse

Every morning, begin by cleansing your skin. Find a cleanser that is appropriate for your skin type. If you have oily skin, look for a foaming cleanser. If your skin is dry, then a hydrating cleanser will be better. For sensitive skin, a gentle cleanser can do the job without causing irritation.

Morning, Step 2: Protect with Antioxidants

Antioxidants fight free radicals, which can damage and age your skin. The most popular and easy-to-find topical antioxidant is vitamin C. Many skin care companies have their own vitamin C serum. Find one and make sure it's packaged in a bottle that doesn't allow light in. Sunlight can oxidize vitamin C, turning it dark brown and rendering it potentially useless.

Morning, Step 3: Protect with a Sunblock

Although antioxidants fight free radicals, they won't prevent ultraviolet rays from harming your skin. The American Academy of Dermatology recommends applying a broad-spectrum sunscreen or sunblock with an SPF of at least 30 to your face every morning. It will block 97% of the sun's rays, preventing your skin from prematurely aging as well as protecting it from future skin cancers.

Evening, Step 1: Cleanse

I can't emphasize enough how important it is to cleanse your facial skin every night to remove the day's buildup of dust, dirt, grime, pollution, oil (sebum), and leftover makeup. If left overnight, these substances will clog your pores and damage your skin. Your facial skin needs to breathe and rejuvenate at night, and it can't do that while covered with dirt and oil.

Evening, Step 2: Treat with an Anti-Aging Cream

By far, the most scientifically studied, clinically proven, anti-aging skin care ingredients are retinoids. To recap, retinoids come in prescription strength (tretinoin, aka Retin-A) and nonprescription strength (retinol). Studies have shown that prescription-strength retinoids can tighten the skin, improve fine lines, lighten pigmentation, thicken the deep layer of the skin, and even reverse early pre–skin cancers.[169] Although you must obtain tretinoin via a prescription or from a doctor's office, there are many inexpensive, over-the-counter skin care products that contain retinol.

Evening, Step 3: Moisturize (Optional)

Many of us like to apply a soothing moisturizer at night. I think this habit goes back to the days of our mothers and grandmothers who applied cold cream to their faces before turning in. If you have dry skin and like to soothe and moisturize at night, by all means go ahead. Pick a moisturizer that won't clog your pores and ideally even has some anti-aging ingredients built-in, like growth factors, peptides, or antioxidants.

ONE TO THREE TIMES WEEKLY: EXFOLIATE

When we are young, our skin cells turn over every six to eight weeks. As we age, this process slows down. It begins to take 10, 12, even 14 weeks for our skin to turn over. This results in skin that is rough, wrinkled, and drab.

Exfoliating your skin regularly can rev up this process. By removing the upper layer of dead skin cells, exfoliation sends a cellular signal to the deeper skin cells to turn over more quickly. This causes skin to look smoother, more radiant, and less wrinkled. Most people with a normal skin type should exfoliate in the evening two to three times per week. If you have sensitive skin, then you may only need to exfoliate once per week.

Exfoliation can be performed mechanically using an exfoliating scrub or chemically with an alpha hydroxy acid at-home mask or peel.

THE ULTIMATE
YOUNGER FOR LIFE SKIN CARE ROUTINE

My Two Minutes, Five Years Younger Routine is the most popular and the one most of my patients and followers prefer, but if you really want to get serious with your anti-aging interventions and maximize your autojuvenation, you can do even more. It will take a little longer and cost a little more, but if you are a tried-and-true skin care enthusiast who wants to do everything possible to keep your skin looking healthy and youthful, this may be the regimen for you.

This routine includes everything in the Two Minutes, Five Years Younger Skin Care Routine but adds the following steps below, made for the person who wants to take the utmost care of their skin.

DAILY
Ultimate Autojuvenation Morning:

1. Cleanse.
2. Tone.
3. Protect with antioxidant serum, ideally one that combines both vitamins C and E for a synergistic effect.
4. Moisturize (optional).
5. Apply eye cream. Pick one that is moisturizing for the sensitive eyelid skin.
6. Protect with sunscreen.

Ultimate Autojuvenation Evening:

1. Cleanse. Consider a double cleanse starting with an oil-based cleanser to remove makeup, followed by your usual cleanser.
2. Tone.
3. Treat with retinol, growth factors, and/or peptides. If you want to do the absolute most for your skin, in terms of anti-aging creams, then combining retinol with growth factors and/or peptides is the way to go. That way you get a one-two-three punch in attacking aging from many different routes. A cream or serum containing bakuchiol is another great option.

4. Moisturize (optional).
5. Apply brightening cream (optional, if you have spots you want to lighten).
6. Apply eye cream, ideally one with low-dose retinol in order to thicken the eyelid skin which can become thinner and crepey with age.

ONE TO THREE TIMES WEEKLY

Exfoliate, once a week for sensitive skin, two or three times per week for nonsensitive skin.

Please check out Appendix 2 for my Dr. Youn–Approved skin care company recommendations and Appendix 3 for my own YOUN Beauty skin care products.

PART FOUR:

THE AUTOJUVENATION JUMP START

CHAPTER FIFTEEN:

THE THREE-WEEK AUTOJUVENATION JUMP START

Even if you've already put many of the concepts I've explained in detail to you about the Younger for Life Diet, the Younger for Life Supplement Protocol, and the Ultimate Younger for Life Skin Care Routine into practice, I've found that people do better, and see results more quickly, when they have a more organized and systematic approach to lifestyle changes. That's why I created the Three-Week Autojuvenation Jump Start. Over the course of three weeks, I'll tell you exactly what to do to begin shifting what's going on inside your body in ways that will help you to feel more energized, look healthier and younger, and even lose a few unwanted pounds.

This program jump-starts the autojuvenating process by flooding your body with collagen-supporting proteins, free-radical-neutralizing antioxidants, anti-inflammatory foods, microbiome-supportive foods, and an intermittent-fasting and ketogenic-inspired phase to ramp up autophagy, that process of cellular cleansing so critical for the reversal of aging. You'll enhance your body's nutritional status with a targeted mix of nutritional supplements, and you'll support your skin with a simple skin care routine that could take years off your face.

At the end of three short weeks, you'll look younger, with improved vibrancy in your skin. You'll begin to fade age spots, lose unwanted pounds, increase your energy, and maybe even reset your metabolism. This is a great starting point for anyone who wants to clean out their system and enjoy noticeable autojuvenative changes in a short period of time.

A big part of the Jump Start—and of the Younger for Life protocol overall—is diet. Specifically for this Jump Start, the diet is divided into two phases. Phase 1 focuses on what to eat to support your collagen—the feasting part of the Jump Start, based on the foods I recommended in the first part of this book. Phase 2 focuses on how to increase autophagy, or the elimination of old cells and cell components, to make room for fresh new cells—the fasting part of the Jump Start based on the section in the first part of this book on autophagy. Phase 2 includes short periods of time where you will eat the Autophagy Diet, consisting of foods that help promote and prolong autophagy. You've already learned all about how and what to eat to autojuvenate, but in this process, I'll organize it for you so you always know exactly what to do.

The Jump Start consists of four parts and includes recipes for every meal and snack.

PART ONE:
THE YOUNGER FOR LIFE DIET, PHASE 1–*FEASTING*

The priorities during this first phase are to:

1. **Reverse collagen degradation**, to prevent skin from getting thinner and to rebuild degraded collagen.
2. **Neutralize free radicals**—particularly, reactive oxygen

species—that cause tissue damage and destroy healthy cells, to reduce oxidative stress.

3. **Douse inflammation**, which not only contributes to the chronic diseases associated with aging but that causes blotchy redness on the skin and contributes to inflammatory skin conditions.

4. **Feed the microbiome**, which is the collection of trillions of bacteria and fungi in the digestive tract. These gut florae have many beneficial effects, from healthy digestion to mood regulation, and they're also instrumental in immunity. A healthy gut equals a vibrantly healthy body and healthier, younger-looking skin. The components of this phase of the diet include:

   • **Collagen-building foods:**
     ° *For nonvegetarians:* Organic, grass-fed or pastured (if affordable and attainable) meat and poultry in small amounts (you don't need much), wild-caught fish, and collagen-rich bone broth.
     ° *For vegetarians:* Organic, non-GMO (if affordable and attainable) soy products, fresh or fermented, such as tofu, tempeh, and miso.
     ° *For all:* Nuts, seeds, and legumes.
     ° *Beta-carotene-rich foods:* Beta-carotene helps support the production of collagen and glycosaminoglycans that improve the skin's ability to retain moisture. The vegetables and fruits that are orange, yellow, or dark green all contain beta-carotene.

   • **Microbiome-boosting foods**, such as kimchi (a personal favorite of mine), kombucha, kefir, sauerkraut, miso, tempeh, and unsweetened yogurt.

   • **The most potent antioxidant-rich foods:**
     ° *Fruits and vegetables* (organic when possible) rich in vitamin C, carotenoids, and polyphenols, for antioxidant action.

° *Herbs and spices*, which are some of the most antioxidant-packed foods available.
° *Dark chocolate* with at least 70% cacao for maximum antioxidants and minimum sugar.
- **Healthy fats**, especially omega-3 fatty acids in fatty fish, oysters, seaweed, and flaxseed, and monounsaturated fatty acids like those in walnuts, olives, and almonds.
- **Green tea/matcha**, possibly the most powerful of all the antioxidant-containing beverages.
- **Black coffee in moderation**, if you already drink it. Coffee is full of antioxidants and is associated with a lower risk of heart disease, dementia, certain cancers, type 2 diabetes, and other diseases associated with aging.
- **Red wine in moderation**, if you already drink it. Small amounts of red wine—like one small glass per day—appears to have a beneficial effect on inflammation and aging, probably because of a polyphenol called resveratrol.
- **Adequate hydration** with clean, pure water.

Note: this phase of the diet contains no refined grains (such as anything made with white flour) or sugar.

PART TWO:
THE AUTOPHAGY DIET, PHASE 2–*FASTING*

The second phase of the diet portion promotes autophagy. This is not a mode to be in all the time, but alternating between Phases 1 and 2 creates the perfect balance of nourishment and cleanup.

During Phase 2, you will practice a 16:8 intermittent fast. You'll fast for 16 hours overnight and restrict eating to an 8-hour window. For example, if you finish dinner by 8:00 p.m., you should not eat anything containing calories until the next day at noon. You can design the schedule in whatever way works for you.

In addition, during this eating window, you will eat the Autophagy Diet, consisting of autophagy-promoting foods. This is similar to a ketogenic diet. These include:

- **Reduced protein.** 25 grams or less in a day. Only do this on Phase 2 days, as you need protein for collagen production. The protein you eat should ideally be grass-fed, pastured, or wild-caught, if it's affordable and accessible.
- **More fat.** Fat has a limited effect on autophagy, so Phase 2 days will be your high-fat days. Emphasize healthy fats: avocados, olive oil, ghee (clarified butter), grass-fed butter, nuts and seeds, coconut oil or MCT oil, and omega-3 fatty acids from fatty fish.
- **More polyphenol-rich foods.** Polyphenols encourage autophagy. Best sources include black beans, plums, cherries, all berries, dark leafy greens like spinach and kale, brussels sprouts, artichokes, red onions, reishi mushrooms, nuts, seeds, ginger, turmeric, black coffee, green tea, Earl Grey tea, and red wine.

Foods to avoid in both phases include:

- **All refined grains.** Anything made with white flour or that has had the bran and germ removed or is otherwise processed, like white bread, white rice, white pasta, and instant oatmeal.
- **Gluten-containing foods.** Drastically limit or preferably discontinue these. You don't have to be super strict if you don't have celiac disease, but give this a try and notice whether you feel better and this improves your skin.

- **All concentrated sweeteners.** Sugar, honey, maple syrup, fruit juice concentrate, agave nectar, etc., and especially sugary drinks (including diet soda), juices, energy drinks, and beer. Ideally, avoid all alcoholic drinks (except red wine, as described earlier).
- **Dairy products.** Eliminate all dairy except kefir and yogurt, which are fermented. Consume these sparingly or use plant-based versions such as almond or coconut yogurt.
- **All refined vegetable oils and trans fats.** Canola, corn, soybean, sunflower, etc., including any processed foods that contain them. Olive oil and avocado oil are fine.
- **Meat from industrial feedlots and farmed fish (non-organic).** Instead, choose organic/pastured/wild-caught meat and fish when possible and affordable. I know not everyone can attain or afford to pay extra for organic, so if it's not possible, then don't stress about it. Try to keep your portions on the smaller side, especially on fasting days.
- **Any processed and fast food.** Packaged snack foods, cured meat, and anything you could get in a drive-through.
- **Any foods containing salt you don't add yourself.** Go easy on the salt.
- **Burnt or charred foods.** These create heterocyclic amines that create advanced glycation end products, which are proinflammatory and speed up the aging process.

PART THREE:
THE YOUNGER FOR LIFE SUPPLEMENT PROTOCOL

Supplements support the body when you don't get everything you need from food. They are an insurance policy against aging and can help make up for days when you don't get enough nutrition, as well as for the reduced nutrition in foods grown in depleted soil. On the Autojuvenation Jump Start, you will be taking these supplements daily:

- **Protein powder.** One scoop per day added to your smoothie, ideally made from pea protein.
- **Hydrolyzed collagen peptides.** One scoop per day added to your smoothie (or a comparable vegan/vegetarian option).
- **A high-quality multivitamin.** Ideally containing the vitamins and nutrients outlined in chapter 4.
- **Fish oil.** Take each day as directed by the manufacturer. If you are vegan or vegetarian, then high-quality algae oil (which is a great source of omega-3 fatty acids) can be a substitute.
- **Probiotic.** Take each day as directed by the manufacturer.
- **A high-quality antioxidant blend.** Take each day as directed by the manufacturer. Ideally pick one with a wide range of antioxidants, such as resveratrol, curcumin, and quercetin.

PART FOUR:
THE YOUNGER FOR LIFE SKIN CARE ROUTINE

During the Jump Start, you'll master the Two Minutes, Five Years Younger Skin Care Routine, as described on page 55. (I'll put it into the schedule below, so you don't have to keep flipping through the book.) I suggest making this a habit to last the rest of your life.

THE SCHEDULE

Go shopping before you begin, based on the lists of recommended foods or the recipes, if you choose to use those exclusively—an easy way to manage the diet component. I've also created a printable shopping list that you can take to the store and a Younger for Life Companion Recipe Book, both of which you can download for free at autojuvenation.com.

Make sure you have all the necessary supplements and skin

care products. I also recommend keeping a written record of everything you're doing and how you feel each day, so you can keep track of your progress.

I encourage before-and-after pics. Take a photo of your face and a full-body photo before you begin the Jump Start, and again three weeks later. Make sure to take the photos in good, natural light and at the same time of day, to keep them standardized. I think you're going to like what you see! The people I've taken through this protocol had noticeable differences, both in their body shape and the youthful look of their skin. But keep in mind, it's a jump start, not a facelift, so make sure your expectations regarding how much autojuvenation you see are realistic.

Here's what you'll be doing during each of the three weeks.

WEEK ONE:

Step One: Eat the Younger for Life Diet on all seven days

Breakfast:
Each day during Week One, choose one of the breakfast smoothies starting on page 197:
• Peachy Green Smoothie
• Blueberry Bliss Morning Smoothie

Lunch:
Each day during Week One, choose one of the lunches starting on page 199:
• Veggie Frittata Muffins
• Black Bean and Sweet Potato Chili (with optional grass-fed ground beef)
• Hearty Miso and Vegetable Soup (with Optional Baked Chicken)
• Oven-Roasted Vegetable Pasta (with Optional Baked Chicken)
• Lentil Vegetable Stew

Dinner:

Each day during Week One, choose one of the dinners starting on page 209:

- Stir-Fried Brown Rice with Veggies
- Grass-Fed Beef or Tempeh Tacos
- Shepherd's Pie with Potato Crust
- Oven-Baked Salmon with Red Onions, Potatoes, and Crispy Kale
- Sheet Pan Chicken with Roasted Vegetables and Kale

Snacks:

Each day during Week One, you can have one of the snacks starting on page 218, either in the midmorning or midafternoon, depending on when you tend to get most hungry between meals:

- Golden Milk Chia Pudding
- Cinnamon-Spiced Overnight Oats with Blueberries

Desserts:

Each day during Week One, you can have one of the desserts starting on page 221, either after lunch or after dinner:

- Dark Chocolate and Coconut Truffles with Orange Zest
- Mixed Berry Cobbler with Gluten-Free Oat/Almond Topping
- Gluten-Free Lemon Blueberry Muffins

Step Two: Follow the Younger for Life Supplement Protocol

- Hydrolyzed collagen peptides: one scoop per day in your smoothie (you can use your favorite brand or my YOUN Beauty Supplemental Collagen)
- Collagen-supporting multivitamin: take as directed by the manufacturer
- Fish oil or algae oil: take as directed by the manufacturer

- Probiotic: take as directed by the manufacturer
- Antioxidant supplement blend: take as directed by the manufacturer

Step Three: Follow the Two Minutes, Five Years Younger Skin Care Routine

Morning: 3 steps
1. Cleanse with an appropriate cleanser
2. Antioxidant protection: apply a vitamin C antioxidant serum
3. Sunscreen protection: apply a quality facial sunscreen

Night: 3 steps
1. Cleanse: use the same cleanser you use in the morning. Double cleanse with an oil-based cleanser to remove makeup if needed
2. Treat: apply a retinol-based moisturizer
3. Moisturize: apply other moisturizer as needed or desired, such as with peptides
4. Twice per week: exfoliate with a chemical exfoliator or a physical exfoliating scrub

Optional: brightening cream, if you have age spots.

WEEK TWO:

This week, you'll be eating differently each day. Follow this schedule.

Day 1: Eat the Younger for Life Diet (as in Week One) but don't eat anything after 8:00 p.m.

Day 2: Don't eat until noon. You may have water, coffee, or tea (preferably green tea), but with no cream or sugar. After noon, eat the Autophagy Diet. On Day 2, you can choose from the following autophagy-promoting lunches and dinners (recipes beginning on page 225).

Autophagy-Promoting Recipes for Lunch and Dinner:
- Creamy Roasted Cauliflower Soup
- Oven-Baked Fish Tacos in Lettuce or Cabbage Leaf Wrap
- Oven-Roasted Vegetables with Pistachio Pesto (with Optional Fish)
- Salmon Burgers in Lettuce Wrap with Miso Mayo, Avocado, and Sauerkraut
- Blueberry and Roasted Almond Salad with Lemony Vinaigrette

Day 3: Eat the Younger for Life Diet, no fasting restrictions, using the recipes from Week One.

Day 4: Eat the Younger for Life Diet, but stop eating by 8:00 p.m., using the recipes from Week One.

Day 5: Don't eat until noon. Eat the Autophagy Diet for the rest of the day, using the recipes from Week Two, Day Two.

Day 6: Eat the Younger for Life Diet, using the recipes from Week One.

Day 7: Eat the Younger for Life Diet, using the recipes from Week One.

On all days, follow the Younger for Life Supplement Protocol and the Two Minutes, Five Years Younger Skin Care Routine from Week One.

WEEK THREE:

Repeat Week Two exactly.

This is a program you can repeat two to three times per year for maintenance. Some may even like to do it quarterly.

OVERVIEW OF THE JUMP START

WEEK ONE:

	SUNDAY	MONDAY	TUESDAY	WEDNESDAY	THURSDAY	FRIDAY	SATURDAY
	BREAKFAST: Peachy Green or Blueberry Bliss Smoothie	**BREAKFAST:** Peachy Green or Blueberry Bliss Smoothie	**BREAKFAST:** Peachy Green or Blueberry Bliss Smoothie	**BREAKFAST:** Peachy Green or Blueberry Bliss Smoothie	**BREAKFAST:** Peachy Green or Blueberry Bliss Smoothie	**BREAKFAST:** Peachy Green or Blueberry Bliss Smoothie	**BREAKFAST:** Peachy Green or Blueberry Bliss Smoothie
	LUNCH: Younger for Life Diet	**LUNCH:** Younger for Life Diet	**LUNCH:** Younger for Life Diet	**LUNCH:** Younger for Life Diet	**LUNCH:** Younger for Life Diet	**LUNCH:** Younger for Life Diet	**LUNCH:** Younger for Life Diet
	DINNER: Younger for Life Diet	**DINNER:** Younger for Life Diet	**DINNER:** Younger for Life Diet	**DINNER:** Younger for Life Diet	**DINNER:** Younger for Life Diet	**DINNER:** Younger for Life Diet	**DINNER:** Younger for Life Diet

WEEKS TWO AND THREE:

	SUNDAY	MONDAY	TUESDAY	WEDNESDAY	THURSDAY	FRIDAY	SATURDAY
	BREAKFAST: Peachy Green or Blueberry Bliss Smoothie	BREAKFAST: Fast—Coffee, Tea, Water is okay	BREAKFAST: Peachy Green or Blueberry Bliss Smoothie	BREAKFAST: Peachy Green or Blueberry Bliss Smoothie	BREAKFAST: Fast—Coffee, Tea, Water is okay	BREAKFAST: Peachy Green or Blueberry Bliss Smoothie	BREAKFAST: Peachy Green or Blueberry Bliss Smoothie
	LUNCH: Younger for Life Diet	LUNCH: Autophagy Diet	LUNCH: Younger for Life Diet	LUNCH: Younger for Life Diet	LUNCH: Autophagy Diet	LUNCH: Younger for Life Diet	LUNCH: Younger for Life Diet
	DINNER: Younger for Life Diet, nothing to eat after 8:00 p.m.	DINNER: Autophagy Diet	DINNER: Younger for Life Diet	DINNER: Younger for Life Diet, nothing to eat after 8:00 p.m.	DINNER: Autophagy Diet	DINNER: Younger for Life Diet	DINNER: Younger for Life Diet

*On ALL days, follow the Younger for Life Supplement Protocol and the Two Minutes, Five Years Younger Skin Care Routine.

CHAPTER SIXTEEN:

JUMP START RECIPES

You can eat anything from the Younger for Life Diet food list while on the Jump Start, but for maximum effectiveness and convenience, follow these recipes. You can be sure you will be compliant with the food list, and I think you're going to love these amazing recipes. It's hard to feel deprived when you get to eat like this.

This chapter contains breakfasts, lunches, dinners, snacks, and desserts for Phase 1 that will maximize all the benefits of the Younger for Life Diet, like antioxidants, anti-inflammatory foods, and complete nutrition. After that, you'll find the Autophagy Diet recipes you can use for lunch and dinner (mix and match) for Phase 2, during which time you'll also be practicing intermittent fasting to promote autophagy.

Looking for more tasty recipes that work with the Jump Start or for healthy and delicious eating every day? I've compiled a Younger for Life Companion Recipe Book with a ton more recipes which you can download and print out for free at autojuvenation.com.

PHASE 1: BREAKFAST SMOOTHIES

Peachy Green Smoothie
Serves 1

Ingredients:

1 scoop YOUN Beauty Complete Beauty Protein Powder
(or your favorite protein powder)
1 scoop YOUN Beauty Supplemental Collagen
(or your favorite hydrolyzed collagen or collagen peptides)
12 oz. unsweetened coconut milk (or almond milk)
¾ cup frozen peaches
1 handful chopped spinach
2 teaspoons almond butter

Preparation:

Place all ingredients in a high-speed blender and blend until smooth. If it's too thick, add some cold water or ice.

Blueberry Bliss Morning Smoothie
Serves 1

Ingredients:

*1 scoop YOUN Beauty Complete Beauty Protein Powder
(or your favorite protein powder)
1 scoop YOUN Beauty Supplemental Collagen
(or your favorite hydrolyzed collagen or collagen peptides)
12 oz. unsweetened coconut milk (or almond milk)
¾ cup frozen blueberries
2 teaspoons almond butter
1 handful chopped kale, chard, or spinach*

Preparation:

Place all ingredients in a high-speed blender and blend until smooth. If it's too thick, add some cold water or ice.

Veggie Frittata Muffins
Makes 6 muffins

Ingredients:

1 cup broccoli florets, chopped
½ cup red pepper, chopped
½ small red onion, chopped
1 tablespoon grass-fed butter, olive oil, or avocado oil
Salt and pepper to taste
5 eggs
2 tablespoons water
½ teaspoon salt

Preparation:

1. Preheat oven to 350°F. Lightly grease 6 muffin tins with oil of choice.
2. Heat a large pan over medium heat. Add the butter or oil. Add onions and cook until soft and translucent. Add the chopped red pepper and broccoli and stir continuously, cooking for 3 to 5 minutes or until they start to soften, then season with a pinch of salt and pepper.
3. Divide vegetables evenly into the 6 greased muffin tins and set aside.
4. Whisk eggs, salt, and water together well and pour over

veggies, distributing evenly (about ¼ cup per muffin tin). Sprinkle with salt and pepper.

5. Bake for 20 to 25 minutes or until cooked through and center begins to brown. Allow to cool a few minutes before removing from muffin pan, then serve warm.

Black Bean and Sweet Potato Chili
(with optional grass-fed ground beef)
Makes 2 or 3 servings

Ingredients:

1 tablespoon avocado or olive oil
½ small red onion, chopped
½ teaspoon garlic powder
½ teaspoon paprika
½ tablespoon chili powder
½ teaspoon dried oregano
½ teaspoon ground cumin
¼ teaspoon sea salt
½ medium sweet potato, peeled and cut into bite-size pieces
1 16-oz. can of black beans, drained and rinsed
1 16-oz. can of diced tomatoes
1 ½ cups vegetable or chicken stock
2 to 3 cups kale or chard, finely chopped
Juice of ½ medium lime
½ teaspoon sea salt
Chopped fresh cilantro, avocado, lime (optional)
¼ pound grass-fed ground beef (optional)

Preparation:

1. Heat oil on medium heat in a medium-size pot.
2. Add red onion and a pinch of salt and allow them to cook until soft and translucent, about 3 minutes.
3. Add garlic powder, paprika, chili powder, oregano, cumin, and 1/4 teaspoon sea salt. Stir until combined. If you are

adding ground beef, add it now and stir to coat, cooking 3 to 5 minutes until meat is browned.

4. Add sweet potatoes and stir to combine, cooking for another minute or so. Add black beans, tomatoes, stock, and the ½ teaspoon of sea salt.

5. Bring to a boil and reduce heat to a simmer, uncovered, for 15 to 20 minutes or until sweet potatoes are fork-tender.

6. Add kale/chard and stir into mixture and allow chili to simmer for another 2 to 3 minutes or until kale is bright green and soft. Remove from heat and add lime juice. Adjust salt to taste, add a sprinkle of pepper, and top with chopped red onion, cilantro, and avocado, adding an extra squeeze of lime if desired.

Hearty Miso and Vegetable Soup
(with Optional Baked Chicken)
Makes 2 or 3 servings

Ingredients:

1 tablespoon olive or avocado oil
½ yellow onion, cut into half-moons
1 large carrot, peeled and cut into matchsticks
5 to 10 shitake mushrooms, sliced
(or any other mushroom sliced to equal 2 cups)
3 cups green kale, chopped
4 cups chicken or vegetable stock
2 tablespoons miso (chickpea or mellow white)
Black pepper and salt to taste
Sliced green onions (optional)

Preparation:

1. Heat olive oil in a large soup pot on medium heat.
2. Add onions and a pinch of salt and stir to coat. Let onions cook on medium for 10 to 15 minutes, stirring occasionally.
3. Add carrots and mushrooms and sauté another 5 minutes.
4. Add the stock and simmer for 10 minutes. Turn off heat and add kale. Cover and allow to sit for 3 minutes.
5. Remove ½ cup of the stock from the pot and place it in a bowl with the miso paste. Whisk to combine thoroughly into a smooth liquid, then add this back to the pot and stir.
6. Serve topped with fresh sliced green onions and more salt and pepper to taste.

Note: if you use low-sodium or no-salt-added stock, you can add a teaspoon or two more of miso or sea salt to taste.

Baked Chicken
Makes enough for Hearty Miso and
Vegetable Soup (page 203), or serves 1 for a meal

Ingredients:

1 medium-size, bone-in chicken breast
Sea salt
Pepper

Preparation:

1. Preheat oven to 350°F.
2. Place chicken breast in glass baking dish and sprinkle generously with sea salt and pepper. Bake in oven for 1 hour.
3. Remove and allow to cool a bit before removing meat from the bone. Cut into bite-size pieces and add to soup.

Oven-Roasted Vegetable Pasta
(with Optional Baked Chicken)
Serves 2 to 3

Ingredients:

½ red onion, cut into half-moons

1 medium carrot, sliced in thin rounds

2 cups cauliflower florets, chopped

10 cherry tomatoes

1 medium zucchini, cut into quarters

10 to 15 kalamata olives, pitted and sliced in half lengthwise

4 tablespoons olive oil (2 tablespoons per tray)

2 teaspoons dried thyme (1 teaspoon per tray)

1 teaspoon sea salt (½ teaspoon per tray)

2 to 3 cups gluten-free penne or fusilli pasta

1 chicken breast, oven-roasted and cut into bite-size pieces

Dressing:

3 tablespoons olive oil

1 tablespoon red wine vinegar

1 tablespoon balsamic vinegar

1 garlic clove, minced

½ teaspoon prepared Dijon mustard

1 teaspoon tamari

½ teaspoon maple syrup

¼ teaspoon salt

Dash of pepper

¼ cup fresh basil, chopped (if available)

Preparation:

1. Preheat oven to 400°F and line two baking sheets with parchment paper.
2. Add carrots, cauliflower, and red onion to one sheet. Drizzle with 2 tablespoons olive oil, 1 teaspoon dried thyme, and ½ teaspoon sea salt.
3. Add cherry tomatoes, zucchini, and kalamata olives to the other pan. Drizzle vegetables with 2 tablespoons olive oil, 1 teaspoon dried thyme and ½ teaspoon salt.
4. Place carrot/cauliflower/red onion pan in oven and roast for 15 minutes, then add the zucchini/kalamata olive/tomato mixture pan to the oven along with it on a separate rack.
5. Continue roasting both pans in oven for an additional 30 minutes.
6. Cook pasta according to directions while vegetables are roasting, drain, and set aside.
7. Whisk dressing in a large bowl and set aside.
8. When vegetables are done, remove from oven and allow to cool for 2 to 3 minutes, then add all vegetables to dressing bowl and stir to coat.
9. Transfer pasta to bowl with vegetables and toss to coat.
10. Season with salt and pepper to taste and add extra olives, herbs, etc., and olive oil as needed.
11. Add chicken breast on top if desired. (See page 204 for directions.)

★This dish is best served warm immediately after cooking.

Lentil Vegetable Stew
Serves 3 to 4

Ingredients:

2 tablespoons avocado oil
½ medium yellow onion, chopped
2 garlic cloves, minced
¼ teaspoon sea salt
3 carrots, peeled and diced
2 celery stalks, diced
2 cups red potatoes, cubed
6 cups chicken or vegetable stock
1 tablespoon chopped fresh rosemary or 2 teaspoons dried
1 cup French green lentils, rinsed well and strained
2 cups hearty greens (kale, chard, spinach), well chopped
1 tablespoon red wine vinegar
Salt and pepper to taste

Preparation:

1. Heat a large soup pot on medium heat and add the avocado oil.
2. Once pot is hot, add onions and cook until soft and translucent, then add the garlic and stir to combine. Cook for 3 to 5 more minutes, stirring frequently to make sure garlic does not brown.
3. Add celery, carrots, potatoes, rosemary, and a pinch of salt to the pot and stir to combine. Cook for an additional 5 to 7 minutes.
4. Add stock and rinsed lentils. Bring to a boil, then reduce

heat to a simmer and cook uncovered for 50 minutes to 1 hour or until lentils are soft.

5. Turn off heat, add vinegar and chopped greens, and stir into stew. Cover pot and allow to sit for 3 to 4 minutes to allow greens to soften.

6. Add salt and pepper to taste. Serve with brown rice, quinoa, or cauliflower rice.

Stir-Fried Brown Rice with Veggies
Serves 2 to 3

Ingredients:

1 cup of cooked brown rice
1 egg, whisked
1 tablespoon avocado oil
½ medium yellow onion, chopped
1 teaspoon garlic, minced
2 cups mushrooms, roughly chopped
1 small carrot, chopped into small cubes
1 cup broccoli florets, chopped
1 teaspoon grated fresh ginger
3 tablespoons tamari
Chopped green onions, cilantro, toasted sesame oil (optional)
1 cup cooked pastured chicken or tempeh, chopped (optional)

Preparation:

1. Heat avocado oil in a large (preferably nonstick) pan on medium heat.
2. Add whisked egg and cook through, breaking cooked egg into small pieces, then set aside in a bowl, leaving pan clean.
3. Add a bit more oil to the pan along with the chopped onion. Cook until soft and translucent.

4. Add garlic, mushrooms, and a pinch of salt and mix well. Cook for 3 to 5 minutes until mushrooms begin to soften.
5. Add carrots and broccoli and cook until soft, 5 to 7 minutes.
6. Push veggies to one side and add cold or cooled rice to the pan along with a tiny bit more oil.
7. Add ginger and tamari on top of the rice, then mix all of the ingredients together with the veggies.
8. Turn heat up a bit to medium-high and stir-fry the rice and veggies until the rice begins to brown a bit, then add the eggs and stir in for another minute or so.
9. At this point you can add already-cooked chicken or tempeh and stir to combine. Cook for a few minutes until heated all the way through.
10. Remove from heat and serve topped with chopped green onions or cilantro. You can drizzle a bit of toasted sesame oil on top to deepen flavor if desired.

Grass-Fed Beef or Tempeh Tacos
Serves 2 to 3

Ingredients:

½–¾ lb. ground beef or an 8 oz. package of tempeh
1 tablespoon avocado oil if using tempeh
(no oil if using ground beef)
2 teaspoons chili powder
½ teaspoon salt
½ teaspoon garlic powder
½ teaspoon cumin
½ teaspoon oregano
1 teaspoon onion powder
Dash of black pepper
½ cup water or stock plus
1 tablespoon tomato paste whisked together
Juice of ½ lime
Chopped cilantro

Preparation:

1. Heat a large skillet on medium-high heat. Add oil if using tempeh.
2. Crumble beef or tempeh into skillet, sprinkle with a bit of salt, and cook until browned, about 7 to 10 minutes. Drain excess grease from pan (if using beef), then return to stove and lower heat to medium. Add all seasonings and stir to coat.
3. Add water or stock/tomato paste mixture. Simmer for 5 to 7 minutes or until liquid is absorbed. Add lime juice and stir into mixture.
4. Serve on a gluten-free tortilla or taco shell (such as Siete brand) topped with fresh chopped lettuce, cilantro, and your favorite salsa.

Shepherd's Pie with Potato Crust
Serves 2 to 3

Ingredients:

1 ½ lbs. yellow potatoes cut into chunks
2 tablespoons ghee, grass-fed butter, or olive oil
½ teaspoon sea salt
Dash of pepper
1 tablespoon ghee, grass-fed butter, or olive oil
½ yellow onion, chopped
1 cup white or cremini mushrooms, roughly chopped
1 ½ medium-size carrots, diced
1 ½ cups kale, chopped
2 teaspoons ghee, grass-fed butter, or olive oil
¾ lb. ground dark turkey meat
1 teaspoon crushed dried rosemary or 2 teaspoons fresh chopped rosemary
2 tablespoons tomato paste
2 tablespoons tamari
1 tablespoon red wine vinegar
Salt and pepper to taste

Preparation:

1. For the "crust," boil the potatoes in a large pot with ½ teaspoon of sea salt for 15 to 20 minutes or until you can easily insert a knife through them.
2. Strain water off and allow potatoes to cool a bit, remove skins if desired and mash with 2 tablespoons ghee/butter/ olive oil, ½ teaspoon sea salt, and a dash of pepper or to taste.
3. Preheat oven to 350°F.
4. Over medium–high heat, in a medium skillet sauté onions and mushrooms in 1 tablespoon of oil until soft.

5. Add carrots along with a dash of rosemary and salt and continue to sauté for an additional 5 to 7 minutes. Add kale and cook until soft. Turn off heat and set aside.

6. In another large pan, heat 2 teaspoons oil of choice and brown the turkey meat with a pinch of salt. Strain juices off if they form.

7. Once turkey is cooked through, add rosemary, tomato paste, tamari, and red wine vinegar. Cook for a couple of minutes, adding salt and pepper to taste. You can add more rosemary, tamari, or vinegar to your liking.

8. Add vegetable mixture to turkey and stir to combine well.

9. Pour mixture into a 9" x 9" glass baking dish.

10. Spread mashed potato "crust" onto the mixture to create a thin layer on top.

11. Bake at 350°F for 35 to 40 minutes.

Oven-Baked Salmon with
Red Onions, Potatoes, and Crispy Kale
Serves 2 to 3

Ingredients:

¾ to 1 lb. salmon filets, skin removed
(most fish counters will remove)
¼ teaspoon sea salt
Pinch of pepper
½ teaspoon smoked paprika
¼ teaspoon garlic powder
4 small yellow or white potatoes, boiled for 15 to 20 minutes
or until fork-tender, cooled, then pre-sliced into
thin ¼–½-inch rounds
3 cups green leafy kale, stem removed, chopped into bite-size pieces
½ small red onion, sliced into half-moons
¼ cup melted ghee, grass-fed butter, or avocado or olive oil
Lemon, salt, and pepper to taste

Preparation:

1. Preheat oven to 400°F. Lightly oil a 9" x 13" glass baking dish with ghee/butter/oil.
2. Wash and pat dry fish and place in oiled pan. Add salt, pepper, paprika, and garlic powder to the salmon.
3. Add the thinly sliced potatoes on top of the salmon and onto open space in the pan. On top of the potatoes place the chopped kale and then the red onions.
4. Pour the ghee/butter/oil all over the fish, potatoes, and red onion. Sprinkle with salt and pepper again.

5. Bake at 400°F for 35 minutes. Serve seasoned with extra salt or pepper if desired and a squeeze of lemon.

*You can boil and cool the potatoes the day before or earlier in the day for this recipe. This makes the recipe quick and easy to throw together.

Sheet Pan Chicken with
Roasted Vegetables and Kale
Serves 2 to 3

Ingredients:

2 medium red potatoes, cubed

2 medium carrots, cut into ½-inch rounds

½ yellow onion, sliced into half-moons

2 tablespoons avocado or olive oil

2 teaspoons garlic, minced

1 teaspoon sea salt

Dash of black pepper

1 teaspoon dried thyme

3 to 4 small or medium-size, bone-in chicken thighs or breasts

1 tablespoon avocado or olive oil

½ teaspoon thyme

¼ teaspoon garlic powder

Sea salt (sprinkled on top of each piece of chicken)

4 cups kale, chopped

1 tablespoon olive oil

Generous pinch of sea salt

Squeeze of lemon (optional)

Chopped fresh parsley (optional)

Preparation:

1. Preheat oven to 425°F and generously oil a baking sheet.
2. In a large bowl, combine potatoes, carrots, onions, garlic, oil, sea salt, pepper, and thyme, and toss to coat all vegetables.
3. Spread vegetable mixture out on sheet pan.

4. Place chicken thighs or breasts on top of vegetable mixture. Drizzle them with remaining oil and sprinkle with thyme and garlic powder plus a sprinkle of salt on each.

5. Roast for 40 to 45 minutes or until internal temperature of chicken is 170° to 175°F.

6. Remove chicken and vegetables from oven and push everything to one side of the pan. Add kale to other side of the pan and drizzle with oil and sea salt.

7. Return to oven for 5 minutes or until kale is wilted, then remove pan from the oven.

8. Move chicken to the side of the sheet pan, and stir the vegetables and kale to combine well.

9. Serve vegetables and chicken topped with a squeeze of lemon and chopped fresh parsley (both optional).

PHASE 1: SNACKS

Golden Milk Chia Pudding
Serves 2 to 3

Ingredients:

1 ¾ cups (1 can) unsweetened coconut milk
¼ teaspoon sea salt
¼ teaspoon vanilla extract
¼ teaspoon cinnamon
½ teaspoon turmeric
½ teaspoon grated fresh ginger
Pinch of black pepper
2 teaspoons honey (more if desired)
⅓ cup chia seeds

Optional toppings:

Toasted coconut flakes, blueberries,
strawberries, or raspberries

Preparation:

1. Place coconut milk, sea salt, vanilla extract, cinnamon, turmeric, fresh ginger, black pepper, and honey in a blender and blend briefly until all ingredients are combined and smooth.
2. Pour ingredients into a medium-size bowl and add the chia seeds. Whisk well until chia seeds are incorporated evenly.

3. Let sit at room temperature for 45 to 60 minutes, whisking once at the halfway point.
4. Stir again and serve as is or with your choice of optional toppings.

Storage: keep in refrigerator for up to 4 days.

Cinnamon-Spiced Overnight Oats
with Blueberries
Serves 2 to 3

Ingredients:

1 ½ cups organic rolled oats
2 ½ to 3 cups coconut or almond milk
½ teaspoon cinnamon
¼ teaspoon vanilla extract
2 tablespoons maple syrup
Pinch of salt
¼–½ cup fresh (organic) blueberries

Preparation:

1. Whisk together the milk, cinnamon, vanilla extract, maple syrup, and salt.
2. Place oats in a glass jar or container.
3. Pour the milk mixture over oats. If needed, add a bit more milk so that oats are covered.
4. Whisk all ingredients together.
5. Place a lid on container and place in refrigerator for four hours or overnight.
6. When ready to eat, remove from refrigerator and stir again. Top with fresh blueberries and serve. You can also heat the oats up prior to serving if you have time.

Dark Chocolate and Coconut Truffles with Orange Zest
Makes 6 to 8 truffles

Ingredients:

¼ cup shredded coconut, plus ½ cup, separated
½ cup plus 1 tablespoon coconut oil
3 tablespoons maple syrup
¼ teaspoon vanilla extract
¼ teaspoon sea salt
2 teaspoons orange zest
½ cup cacao or cocoa powder

Preparation:

1. Using a food processor fitted with an S blade, add shredded coconut, coconut oil, maple syrup, vanilla extract, sea salt, and orange zest. Process until well-combined.
2. Pulse in cacao/cocoa powder until incorporated completely.
3. Sprinkle ½ cup shredded coconut on a medium-size plate.
4. Using a tablespoon, scoop mixture out and roll into balls.
5. Roll chocolate balls in coconut until coated.
6. Place truffles in refrigerator for one hour and serve.
7. Store in refrigerator for up to one week.

Mixed Berry Cobbler with
Gluten-Free Oat/Almond Topping
Serves 4

Ingredients:

Filling:

2 cups blueberries
4 cups strawberries, sliced
1 tablespoon arrowroot powder dissolved in 2 tablespoons water
2 tablespoons coconut palm sugar
Pinch of sea salt

Topping:

1 cup gluten-free rolled oats
½ cup almond flour
2 tablespoons chia seeds
½ cup finely shredded coconut
¼ cup coconut palm sugar
1 teaspoon cinnamon
Pinch of sea salt
⅓ cup coconut oil, gently melted
¼ cup maple syrup

Preparation:

1. Preheat oven to 350°F.
2. Place blueberries and strawberries in a large bowl and add the dissolved arrowroot powder, coconut palm sugar, and sea salt. Stir to coat.
3. Pour mixture into a 9" x 9" glass baking dish.

4. In another large bowl combine the oats, almond flour, chia seeds, shredded coconut, coconut palm sugar, cinnamon, and sea salt. Mix all ingredients well.
5. Add the melted coconut oil and maple syrup to the oat mixture and stir to coat.
6. Gently scoop topping onto fruit mixture and spread out evenly.
7. Bake in oven for 45 minutes.
8. Cool for 15 to 20 minutes before serving or serve completely cooled.

Gluten-Free Lemon Blueberry Muffins
Makes 6 muffins

Ingredients:

1 ¼ cup blanched almond flour
¼ cup tapioca flour
½ teaspoon baking soda
¼ teaspoon sea salt
1 egg
⅓ cup maple syrup
1 tablespoon coconut oil, melted
1 tablespoon lemon juice
½ teaspoon lemon zest
½ teaspoon vanilla extract
½ cup fresh or frozen blueberries

Preparation:

1. Preheat oven to 350°F and line 6 large muffin tins with paper liners.
2. In a large bowl, mix together flours, baking soda, and sea salt.
3. In another medium bowl, whisk together egg, maple syrup, coconut oil, lemon juice, lemon zest, and vanilla extract.
4. Add wet ingredients to dry and stir to combine. Fold in blueberries.
5. Using a ¼-cup measuring cup, scoop batter into muffin tins (should be enough for 6 muffins).
6. Place in oven and bake for 35 minutes until tops are firm to the touch. Remove from oven and allow muffins to cool on a cooling rack. Best served almost completely cooled.

Creamy Roasted Cauliflower Soup
Serves 2 to 3

Ingredients:

4 cups cauliflower florets, cut into 1-inch pieces
2 tablespoons avocado or olive oil
1 teaspoon garlic powder
½ teaspoon sea salt
1 additional tablespoon avocado or olive oil
½ medium yellow onion, chopped
3 cups vegetable or chicken broth
2 tablespoons lemon juice
Pinch of black pepper
Chopped flat-leaf parsley or chives (optional)

Preparation:

1. Preheat oven to 425°F.
2. Line a baking sheet with parchment paper and set aside.
3. Place cauliflower florets in a large bowl. Add oil, garlic powder, and sea salt to the cauliflower and stir to coat.
4. Spread cauliflower out evenly on baking sheet and roast in oven for 25 minutes or until it starts to brown.
5. While cauliflower is roasting, heat a soup pot over medium heat and add additional avocado or olive oil.

6. Stir in onions and allow to caramelize for 10 to 15 minutes. You will know they are done when they turn a brownish color and a have a sweet smell. Turn off heat and set caramelized onions aside.

7. When cauliflower is done, remove from the oven and add it to the soup pot along with the broth. Bring to a boil and then turn heat down, cover pot, and simmer for 15 minutes.

8. After 15 minutes, remove from heat and allow to cool for 5 to 10 minutes. Add soup ingredients to a blender, small amounts at a time, and blend until smooth and then return to pot. Add lemon and pepper and extra salt to taste.

9. Heat up a bit if needed and serve warm, topped with chopped chives, green onions, or cilantro.

Oven-Baked Fish Tacos in
Lettuce or Cabbage Leaf Wrap
Serves 2 to 3

Ingredients:

½ lb. cod filet cut into 1-to-2-inch strips
2 teaspoons avocado oil
¼ teaspoon chili powder
¼ teaspoon sea salt
Pinch of black pepper
¼ teaspoon garlic powder
¼ teaspoon oregano

Preparation:

1. Preheat oven to 375°F.
2. Line a baking sheet with parchment paper.
3. In a small bowl, add fish slices, avocado oil, chili powder, sea salt, pepper, garlic powder, and oregano. Stir to combine. Make sure all fish is evenly coated with spices.
4. Place coated fish on baking sheet leaving space between each piece. Bake for 20 to 25 minutes.
5. Serve in a lettuce leaf or cabbage shell/wrap topped with guacamole (see recipe below), chopped tomato, red onion, salsa, or any other toppings you might like.

Guacamole

Ingredients:

1 medium avocado
Juice of ½ lime
1 tablespoon red onion, chopped
½ small Roma tomato, seeded and chopped
¼ teaspoon sea salt
¼ teaspoon garlic powder
1 to 2 tablespoons cilantro, chopped (optional)

Preparation:

1. Remove pit from avocado and scoop flesh into a medium-size bowl. Mash avocado and lime juice until combined.
2. Add red onion, tomato, sea salt, and garlic powder, and optional cilantro and mix all ingredients until well-combined.

Toppings:

Chopped cilantro
Sliced avocado
Chopped tomato
Diced red onion
Lime juice
Salsa
Chopped lettuce
Chopped cabbage

Oven-Roasted Vegetables with Pistachio Pesto (with Optional Fish)
Serves 2

Ingredients:

¼ medium red onion, sliced into half-moons
2 cups cauliflower florets, cut into 1-inch pieces
2 cups broccoli florets, cut into 1-inch pieces
¼–½ medium red pepper, sliced into matchsticks
Optional: ½–¾ lb. fish filet (salmon or cod)

Dressing:

¾ teaspoon sea salt
½ teaspoon garlic powder
Pinch of pepper
4 tablespoons olive oil
1 teaspoon prepared Dijon mustard
1 tablespoon apple cider vinegar

Pesto:

½ cup shelled raw pistachios (can also use roasted,
lightly salted pistachios)
½ cup basil leaves
½ cup Italian parsley leaves
¼ cup olive oil
Juice of ¼ lemon
½ teaspoon sea salt

Preparation:

1. Preheat oven to 400°F. Line a large baking sheet with parchment paper.
2. Add red onion, cauliflower, broccoli, and red pepper to a medium-size bowl.
3. In a small bowl, whisk together dressing ingredients. If you are adding fish to this recipe, set aside a tablespoon or so of dressing.
4. Add dressing to vegetables and stir to coat. Spread dressed vegetables evenly on baking sheet. If you are adding fish, place on baking sheet along with vegetables and drizzle dressing over it.
5. Roast in oven for 30 to 35 minutes.
6. While vegetables are roasting, fit a food processor with an S blade. Add all pesto ingredients and pulse until a paste is formed. Remove from food processor and set aside.
7. When vegetables (and fish, if used) are done roasting, remove from oven and allow to cool for 2 to 3 minutes. Serve warm, topped with desired amount of pesto.

Salmon Burgers in Lettuce Wrap
with Miso Mayo, Avocado, and Sauerkraut
Serves 4 to 5

Ingredients:

1 lb. salmon filet, skin and bones removed
¼ cup red onion, chopped
1 teaspoon lemon zest
¼ cup curly parsley, chopped
¾ teaspoon sea salt
1 teaspoon prepared Dijon mustard
1 egg
2 tablespoons avocado oil

Preparation:

1. Cut salmon into large pieces and set aside.
2. Using a food processor fitted with an S blade, pulse the red onion, lemon zest, parsley, and sea salt until minced well. Scrape down sides of processor as needed to ensure the texture is uniform.
3. Add salmon, mustard, and egg and gently pulse until well-combined.
4. Heat a large nonstick frying pan on medium heat and add the avocado oil.
5. Once hot, scoop approximately ¼ cup of salmon mixture onto pan and flatten out into patties.
6. Cook for 3 to 4 minutes per side or until they start to brown. Flip and then cook an additional few minutes, until browned on second side.
7. Serve on lettuce leaf wraps topped with miso mayo (see recipe below), avocado, sauerkraut, and any other toppings you might like.

Miso Mayo

Ingredients:

1 cup cashews, soaked 4 hours
1 tablespoon mellow white miso
½ teaspoon garlic powder
Juice of ½ to 1 lemon (to taste)
¼ teaspoon sea salt
¼ cup water

Preparation:

Add all ingredients to a high-speed blender or food processor fitted with an S blade and blend until smooth. Add small amounts of water if needed to allow mixture to blend smoothly. Add salt or lemon to your liking.

Blueberry and Roasted Almond Salad with Lemony Vinaigrette
Serves 4

Vinaigrette Ingredients:

½ cup olive or avocado oil
3 tablespoons balsamic vinegar
2 tablespoons lemon juice
1 tablespoon lemon zest
1 teaspoon Italian seasoning mix
½ teaspoon onion powder
½ teaspoon sea salt
1 garlic clove, minced
½ teaspoon prepared Dijon mustard

Preparation:

Whisk all ingredients together until smooth.

Roasted Almonds Ingredients:

2 cups almonds
1 tablespoon avocado oil
1 teaspoon sea salt
½ teaspoon garlic powder
½ teaspoon onion powder

Preparation:

1. Preheat oven to 325°F.
2. Line a baking sheet with parchment paper. Spread almonds out on baking sheet.

3. Sprinkle oil, sea salt, garlic powder, and onion powder on almonds and stir to coat evenly. Spread nuts out again evenly on sheet.
4. Roast in oven for 20 minutes. Remove and allow to cool on baking sheet.

Salad Ingredients:

1 head romaine lettuce, chopped
1 pint organic blueberries
½ bunch curly parsley, chopped (optional)
1 bunch red radishes, thinly sliced (optional)

Assembly:

1. Place a handful of romaine, a handful of blueberries, a few almonds, a few radishes, and a tablespoon of chopped parsley in a bowl.
2. Drizzle desired amount of dressing on salad and toss to coat evenly.
3. Serve with extra dressing, more radishes, extra parsley, or more blueberries or nuts if desired.

PART FIVE:

REVITALIZE WITH
AN ANTI-AGING LIFESTYLE

CHAPTER SEVENTEEN:

AUTOJUVENATE WITH BETTER SLEEP

Once you've gotten your diet, supplements, and skin care routines in place, you are probably already noticing some pretty profound changes in how you feel and look. But there's so much more you can do! The way you live your life can accelerate or slow aging, so a true autojuvenating lifestyle will take more into account, and your next priority, if you really want to slow down aging, is to do a better job of getting a good night's sleep.

Sleep is super important as we age. "Nearly every disease killing us in later life has a causal link to lack of sleep," says Dr. Matthew Walker, a UC Berkeley professor of neuroscience and psychology. "We've done a good job of extending life span but a poor job of extending our health span. We now see sleep, and improving sleep, as a new pathway for helping remedy that."[170] Dr. Walker isn't the only one who thinks sleep is a common problem. A *Consumer Reports* survey found that 164 million Americans struggle with sleep at least once a week—that's 68% of the American population.[171] We spend tens of billions of dollars on sleep remedies every year, yet we're still having trouble sleeping. Many Americans are using prescription sleep aids like Ambien and benzodiazepines.

Problem is, the sleep you get on Ambien isn't the same sleep you get naturally.[172] But sleep is critical to reversing aging. One study published in *Clinical and Experimental Dermatology* showed that people who sleep seven to nine hours had better quality skin and looked younger than people who sleep fewer than five hours per night.[173]

So how do you get more sleep—and better quality sleep?

REGULATE YOUR CIRCADIAN RHYTHM

Your circadian rhythm is the internal rhythm of your body governed by biochemical reactions, such as the release of cortisol in the morning to help you wake up and the release of melatonin in the evening to help you fall asleep. Many things influence circadian rhythm, including when you eat, when you're exposed to sunlight (or lights that mimic sunlight), when you're exposed to darkness, and the regularity—or irregularity—of your sleep schedule.

You can begin to regulate your circadian rhythm by setting a sleep schedule. If you always go to bed and wake up at about the same time, your body will learn the pattern and begin to regulate your biochemistry to comply with it. This in turn will help you feel sleepy at the right time, wake up at the right time, and ensure you're getting enough shut-eye.

Let's say you decide on lights-out at 11:00 p.m. and the alarm for 7:00 a.m. That's a solid eight hours. At first, this may require a transitional period. If you are used to staying up later, you may have trouble falling asleep at 11:00 p.m. until your body adjusts, but keep at it, because your body is wise, and it will figure out what you're doing. As your circadian rhythm adjusts and you get more sleep and have more energy, your skin will likely begin to look better, too. You'll feel more alert and more in tune with your body.

That doesn't mean you can't ever stay up late or sleep in. Breaking your routine once in a while won't hurt—what matters is what you're doing *most of the time*. Your body loves routine and will respond well once you start practicing this.

Back during my plastic and general surgery residency years, I used to be awake at all hours. I had to get my sleep whenever I could, whether that was morning, afternoon, evening, or night. Often I would be awake all night in the hospital performing surgeries and treating patients, then finally get back home the next day and conk out for a few hours.

Fortunately, my life no longer resembles that of my residency years, as this is decidedly not a recipe for long-lasting health and youth. My sleep hours have been pretty consistent over the last two decades. I go to bed around 11:00 p.m. and wake up between 6:00 and 7:00 a.m., depending on whether I'm operating that morning. I tend to feel my best when I get between seven and eight hours of sleep a night. Although many people can thrive with less than that, I believe habits like Martha Stewart's four hours of sleep a night guarantee premature aging for most of us. (Although Martha looks fantastic. She has clearly figured out what works for her!)

Honestly, there is only one thing I want to do at 2:00 a.m. every night. No, it's not hang out at the bar with buddies, or watch reruns of *Friends* on Nick at Nite, or tend to an emergency in the ER (heck, no!), or even get busy. The only thing I want to do at 2:00 a.m. is sleep. Do you agree?

INSTITUTE A BEDTIME RITUAL

Another way to cue your body that it's time to feel sleepy is to create a bedtime ritual and do it every night (or at least on most nights) before you go to sleep. This helps to solidify the message to your body that you have a routine, and it will send signals to your brain to begin releasing melatonin and calming down your nervous system so you can get a good night's sleep. Personally, I always brush my teeth, wash my face, apply my skin care creams, and read a bit before lights-out. Since you may be in the process of creating a new skin care routine yourself, this could be (dare I say should be?) part of your bedtime ritual.

EXERCISE DURING THE DAY

There are plenty of good reasons to exercise that I'm sure you already know: exercise builds muscle to combat frailty with aging;[174] increases circulation to the skin;[175] strengthens the heart and lungs;[176] improves mood in ways that can help you to feel happier, more motivated, and younger (even after just one exercise session);[177] and keeps the body fitter, stronger, and less subject to the effects of aging.[178]

Exercise is also valuable for improving sleep. After all, one of the primary purposes of sleep is to rest an active body and heal. If you are lying around all day, you don't have much to rest from. Exercising each day (but not too close to bedtime, which can be overstimulating) can increase the quality of your sleep[179] and make it easier to fall asleep. Getting vigorous exercise during the day is best, but any exercise is better than none. If you're fairly sedentary right now, start gradually and work your way up to more vigorous exercise. It's better to exercise moderately every day than to go too hard, get injured, and be laid-up completely.

SKIP THE SLEEP-COMPROMISING NAPS

Some people can nap without a negative effect on nighttime sleep, and if that's you, then you can skip this one. How napping affects people is individual. For some, taking a short power nap (10 to 20 minutes) during the day may not affect overall sleep and, in fact, can help improve your performance at work. Back in 2015 or 2016, I visited the *Huffington Post* offices, and they had two designated nap rooms. They had a nice reclining chair and couch inside each room, and employees would sign in on a clipboard next to the door to log their nap time. Studies show that allowing employees to take a nap can improve morale and performance,[180] so I also put a couch in my break room and allow my employees to take a short power nap if needed.

However, there are those whose naps keep them from falling asleep easily at night, and if that's you, then skipping the siesta could help to get your sleep schedule and sleep quality back to where you want it to be. If you really do need a nap during the day, try to keep it to 20 minutes or less.

DIAL BACK THE CAFFEINE

If I tell you to skip your morning coffee, you might decide not to listen to anything else I have to tell you, so let me be clear that I'm not saying you can't start your morning with a little bit of your Starbucks dream brew. However, I know people who have coffee after dinner. The half-life of caffeine (the time it takes for 50% of it to be cleared from your system) is five to six hours on average, so if you're going to use caffeine, morning is really the only appropriate time.

Just to be safe, if you want all the stimulating caffeine out of your system by bedtime, you should avoid it at least eight to ten hours prior to sleeping, especially if you have problems falling asleep. Whenever I want to join in on the coffee-after-dinner tradition, I always order decaf. Or better yet (since even decaf has a small amount of caffeine), a nice hot cup of herbal tea.

MAKE YOUR BEDROOM SLEEP CENTRAL

You've probably heard the old advice that your bedroom should be the place for only two things: sleeping and sex. It's good advice. Make your bedroom a place for just these things, focusing on the decor (relaxing), temperature (slightly cool—between 65° and 70°F, ideally), noise level (low to none), and darkness.[181] My wife and I put up black-out curtains in our bedroom and find that we sleep much better. Have you ever slept in a cruise-ship cabin without windows? Amy and I have, and it was so dark that we felt like we could sleep forever. And we did, even missing our excursion one day!

I'm also sensitive to noises and have a hard time falling asleep when my neighbors are loud. If you're like me, then you may benefit from a white-noise machine. We have an air filter that creates anything from a low hum to a jet-engine rumble. With dogs who freak out at the sound of thunder, it's proven to be very useful for giving us all a good night's sleep!

TURN OFF THE SCREENS

We are a screen-centric culture now. There's no denying it. People often lie in bed scrolling on their phones right before turning off the lights, but this is highly disruptive to sleep quality. The general advice is to avoid screens for two hours before bed, and that's all screens: phones, computers, even television. But what the heck are you going to do for two hours? People these days are often at a loss.

I feel this pain. This one is hard for most of us because there is just so much to see, read, watch, and learn on those screens. We have the whole world in the palms of our hands, and now I'm telling you to put it all away and do what—stare at the wall? And don't get me started with getting sucked into TikTok videos or YouTube shorts. An hour can pass in the blink of an eye... or the scroll of a finger.

Here's the problem. Using electronics (like your computer, iPad, tablet, e-reader, or smartphone) exposes you to blue light. This can suppress the secretion of melatonin, the hormone that regulates your circadian rhythm and tells you when to fall asleep.[182] The morning sun emits blue light, so when in the evening the blue light from your screen enters your eyes and hits your pineal gland, your brain thinks it's daytime and tells you to wake up. Inconvenient, but that's what happens. Although any light can suppress melatonin and interfere with circadian rhythm, blue light has the greatest effect.

To their credit, smartphone manufacturers have taken this knowledge into consideration and created the Night Shift mode on your phone. When you turn it on, it can reduce the brighter colors and the blue spectrums, making them warmer and exposing you to less blue light at night. On an iPhone, you can set the Night Shift to turn on every evening and turn off every morning.

Another option is to try blue-light-blocking glasses. If you've ever seen anyone wearing amber, red, or orange-colored lenses in the evening, chances are that blocking blue light for better sleep is the reason. Today these glasses can even have completely clear lenses, so no need to stand out with them on. I bought a pair for each member of my family, and my son tells me he gets less eye strain and headaches when he plays video games with them on, allowing him to game for longer than he would otherwise. If you have children who are gamers, then feel free to keep this information away from them, ha!

You can also simply turn down the lights in the evening so you don't have any bright overhead lights that your brain might mistake for sunlight. All these hacks can certainly help, but also consider that screen use and bright lights can be stimulating in general. Whether you're sitting in a brightly lit room or watching an exciting movie, relaxing can be difficult. If you possibly can, try to turn the lights low and curb your visual-media consumption at least for a short time before getting into bed, and avoid scrolling while in bed, especially if you have a hard time falling asleep. But if you must, I hope you'll turn on the Night Shift mode and/or try out blue-light-blocking glasses in the evenings.

CONSIDER THE POSSIBLE INFLUENCE OF EMFS

Electromagnetic frequencies, or EMFs, are the electromagnetic waves that come from devices like phones, Wi-Fi, computers, and indeed anything you plug into the electricity in your home. This is, I admit, a controversial subject. Some people don't believe EMFs have any effect on humans. However, there are many anecdotal reports, and now some studies, showing that exposure to electromagnetic frequencies, such as from your phone, wireless signal, and other sources, can negatively impact your sleep.[183]

I once stayed at a hotel in downtown Chicago for a conference. The hotel was a converted courthouse and just seemed to have bad energy, which I noticed the moment I walked in. I couldn't find a reason—it was just a strong feeling. Since I arrived late in the evening, they stuck me in the only room still vacant in the hotel. For the first two nights I could not sleep. I would lie awake for hours, and even though I was physically tired, my body and mind would not allow me to drift off. After two nights, I realized that one of the hotel's Wi-Fi boxes, that emits the Wi-Fi signal for everyone in the building, was in the corner of my room! I switched rooms and for the next two nights, I slept like a baby. Coincidence? I think not.

Most previous studies did not seem to show a correlation between exposure to EMFs and sleep disturbances, but these studies were very limited in duration and scope.[184] More recent rat studies show all sorts of abnormalities that can occur with excessive exposure to EMFs. Studies in humans are still limited but need to be done, especially with the omnipresence of 5G and the increasing amount of time we spend on our cell phones and tablets. Just because we don't know for sure whether EMFs are having a negative effect on us doesn't mean they aren't. I'm definitely not a conspiracy theorist, but there have been plenty of things we all thought were totally fine until research proved otherwise (like smoking!).

Until science weighs in with more authority, what are some practical tips to reduce EMF exposure while you sleep? First and foremost, do not sleep on your phone. A lot of people sleep with their phones under their pillows, often using them as an alarm clock in the morning. Don't do it! Unless your phone is in Airplane mode with Wi-Fi and Bluetooth turned off, your phone is regularly emitting signals. Do the same thing with your smart watch. Also do not put your wireless router next to your bed. These are continually sending signals, so best to keep yours as far away from your sleeping space as possible. If you are EMF-sensitive, then you might want to take the extra step of turning your router off before bed.

TRY NATURAL SLEEP AIDS

I'm not a proponent of pharmaceutical sleeping pills because of the side effects, and I think it's healthier all around to avoid unnecessary medication. However, there are some natural sleep aids that seem to work very well. One of them is melatonin. Melatonin is the sleep hormone that your own pituitary gland produces in response to low light and that signals to your brain that it's time to sleep. This should work without supplementation, but disrupted circadian rhythms can lead to disrupted melatonin production, and everyone produces less melatonin as they get older.

Taking melatonin in supplement form is an easy way to help your body fall asleep faster without the morning hangover effect of medications.[185] I often use melatonin as a sleep aid when traveling to other time zones.[186] I recommend taking 1–5 mg (I use 2–4 mg, depending on how much I need it) about 30 minutes before bedtime.

Valerian root, a popular and proven herbal supplement, also

makes a good sleep aid.[187] Unfortunately, when I've taken it, I've had some hangover effects the next morning, so I prefer melatonin over valerian, but it seems to work quite well for some people.

Essential oils have become a trendy way to naturally improve sleep, and they are both inexpensive and highly effective for some. Not only can essential oils help you fall asleep, but they may help sleep be deeper for longer. Some of the aromas that seem to have the best effect are lavender, bergamot, Roman chamomile, and ylang-ylang.

Lavender is one of the most commonly used essential oils to help with sleep. It can reduce feelings of stress and anxiety, calm the nervous system, and even help ease headaches. (Although please be aware that lavender oil can be toxic for pets.) Put lavender oil in a diffuser in your bedroom and let it run for up to two hours when you sleep.

Like lavender, bergamot has a calming and soothing aroma. Studies show it can reduce heart rate and lower anxiety and stress.[188] Roman chamomile also has calming effects. Most of us think of chamomile tea when we hear that herb's name. My mom drinks chamomile tea in the evening due to its calming effects, and sometimes I do, too. Ylang-ylang essential oil has a pleasant, flowery aroma and a calming effect on the nervous system.[189] One study found that combining ylang-ylang with bergamot and lavender reduced stress, anxiety, and blood pressure in people with hypertension.[190] Not bad!

EARTHING/GROUNDING

Did you know that the simple (even primal) act of putting your feet on the bare ground can improve your sleep? In the world of holistic health, this is called *earthing* or *grounding*, and the theory is that exposing your body to the bare earth can draw in electrons which supposedly can act similar to an antioxidant, neutralizing free radicals.[191] One study showed that grounding the human body to the earth during sleep via a grounding mat that is wired into the ground can reduce nighttime levels of cortisol (our body's natural stimulant, also known as the primary stress hormone), help regulate circadian rhythm, and even improve the quality and restfulness of sleep.[192]

In at least one other study, grounding has been shown to reduce inflammation and alter the numbers of circulating white blood cells, cytokines, and other markers of the inflammatory process.[193] While the jury is still out in the eyes of mainstream science, it makes sense to me that we came from the earth and that contact with the earth, especially in a world in which we spend most of our time indoors, is bound to feel like a relief to our systems. Why not give it a try? It's easy and free. Take off your shoes and socks and stand in the dirt, on the grass, or on the sand of a beach whenever you can! It certainly can't hurt, and it might even help.

I hope you've gotten some good ideas and motivation for prioritizing sleep. A well-rested face certainly looks younger, but a well-rested body works better and will be more adept at staying younger longer.

CHAPTER EIGHTEEN:

AUTOJUVENATE BY MANAGING STRESS

When COVID first hit in March 2020, my stress levels hit an all-time high. For the first time since medical school, I found myself waking up at 2:00 or 3:00 a.m., lying awake worrying and unable to fall back asleep. During that time, I felt like I'd aged 10 years.

Stress does age us. You need only look at photos of the US presidents when they entered office and then four or eight years later. Their time in office tends to age them dramatically.

Stress management is one of the most valuable things you can do for yourself because it will affect everything else. You probably have experienced how hard it is to eat healthy meals, take a walk, remember your supplements, take care of your skin, or get a good night's sleep if you're stressed. Stress, or the lack of it, underlies everything, so even though this is at the end of a long list of things you can do to autojuvenate, stress management is as important as any other intervention in this book.

This is why the next most important lifestyle alteration you can make is to manage your stress.

Stress Shortens Your Telomeres

Stress shortens the telomeres on our chromosomes. As you may remember from the beginning of this book, telomeres are protective caps at the end of our DNA strands that allow our cells to replicate and divide. The longer the telomere, the more times a cell can divide. As cells divide and get older, the length of the telomeres gets shorter. The shorter the telomeres, the more susceptible our cells are to dying and the more susceptible we are to disease.

This is a natural aging process that determines the life span of the cells. Shorter telomere lengths are linked to poorly functioning immune systems, cardiovascular disease, osteoporosis, and Alzheimer's disease. Psychological stress in general has been scientifically proven to shorten telomeres,[194] which may be an important reason why stress is associated with both declining health and accelerated aging.

Because meditation is used very successfully to reduce psychological stress, it would make sense that it would also contribute to keeping telomeres longer—and it does! Meditation has been shown in multiple studies to help maintain telomere length.[195] By keeping telomeres longer, meditation can slow down aging and even prolong your life span.

MEDITATION

So how do you manage stress? There are many ways, but my favorite, and the one way that has been well-studied and proven to make an actual physical difference on the body's stress, is meditation.

During COVID, as my wife Amy's and my stress grew chronic,

we decided to try meditation. Sometimes I would meditate on my own without a guide, focusing mainly on my breathing for 20 minutes while sitting in a quiet room with my eyes closed, or sometimes I'd use an app. Although my mind would often wander and I'd find myself worrying, I tried my best to focus on breathing.

Meditation made a real difference for both of us. I found that on those days when I meditated, I slept much better. I would even sleep all night like I had before the pandemic started. On days where I skipped meditating, my sleep would often be disrupted. I realized that the simple act of being still and paying attention to your breathing for 10 or 20 minutes each day could profoundly impact your life. This isn't the only meditation technique, but it's a simple one anyone can do without any training. Just sit down. Breathe. Listen to your breath. Feel it going in and out. That's all it takes.

There's quite a lot of research supporting meditation as well as deep breathing for physical and mental health,[196] stress reduction,[197] and anti-aging,[198] including improving memory loss related to aging[199] and actually slowing down aging at the cellular level.[200] The stress-reducing properties of meditation seem to have a ripple effect throughout the entire body. One study even found that meditation can improve the results of treatment for skin issues like acne and psoriasis![201]

If you would like to try meditation, there are multiple methods. Like I previously mentioned, you could simply sit and focus on your breathing. Deep breathing alone can reverse the stress response,[202] and diaphragmatic breathing in particular can help your body shift into a more relaxed state,[203] reducing the level of the stress hormone cortisol in your blood and making you feel calmer and in a better mood.[204] To do this, focus on expanding your diaphragm—a round muscle under your lungs—as you breathe deeply, by allowing your abdomen to expand and contract without moving your shoulders.

Another way to practice deep breathing in an organized way is with a kind of breathing popularized by Dr. Andrew Weil called 4-7-8 breathing, in which you exhale fully, then inhale through your nose for a slow count of 4, hold for a count of 7, and exhale through your mouth for a count of 8. Different types of deep breathing are fundamental to a yoga practice called pranayama, which is where many deep breathing exercises originated.

If you want to do something beyond deep breathing, you could follow along with a guided meditation (like Amy and I did) from an app or any of the many guided meditations available online for free. You could sit quietly with your eyes closed and imagine calming scenes or being in a favorite relaxing place, or, with your eyes open, focus on a single point, like a candle flame or a picture of a calming scene.

There are virtual-reality apps that help you meditate, or you could meditate while walking slowly and focusing on the beauty of nature. Or you could try mindfulness meditation, during which you focus on all the sense impressions you are getting in the moment, or mantra meditation, where you continually repeat a word or phrase that makes you feel relaxed, such as *peace*, *love*, or *calm*. When I don't have access to a meditation app, then I often will practice mantra meditation.

Regardless of the type of meditation you do, the real benefits come from consistent daily practice (although even intermittent sessions can have benefits, according to one study of users of a meditation app[205]).

YOGA AND STRENGTH TRAINING

Is walking all you need to do for exercise as you get older? Heck, no!

There's a common fallacy that as you age, all you need to do is walk for sufficient exercise. Now, don't misunderstand me. Walking is great and definitely an upgrade over a sedentary life, but it doesn't work all your muscles or help you with strengthening, flexibility, or balance. All of these are necessary for staying more mobile as you age. But walking can only do so much. You'll need other forms of exercise to help with balance and staying limber, strong, and healthy, especially when you enter your fifth decade or more.

Skeletal-muscle fibers come in two types: slow twitch and fast twitch. Slow-twitch muscle fibers are used for endurance and energy. These are the muscle fibers which are strengthened with exercises like walking. Fast-twitch muscle fibers are used to support quick, powerful movements and are strengthened with exercises like weightlifting and sprinting. If all you do is walk, then you're only working out those slow-twitch muscle fibers and muscle groups that are used to propel you forward. You're neglecting fast-twitch fibers, which decline in function with age, and all the *other* slow-twitch muscle fibers which are key to keeping your balance, strength, and endurance.

One of the worst things that can happen to an older person is a fractured hip. In fact, one study found that 1 in 3 adults over the age of 50 dies within 12 months of a hip fracture.[206] That's a frightening statistic. In older adults, hip fractures are most often caused by falling from a standing position, often falling sideways.[207] So how do you prevent falls like this as you get older? Do exercises that ensure your fast-twitch muscle fibers stay strong and active and that keep your balance and core strong.

When you accidentally trip, let's say on a step you don't see, the muscle fibers that suddenly stabilize you and prevent you from falling are the fast-twitch fibers. In the milliseconds after your body becomes unsteady, those muscle fibers spring into action and activate the muscles which prevent you from falling and hurting yourself. So, as you can see, these muscle fibers are absolutely crucial to protecting yourself from injuries, especially as you age. If those muscle fibers aren't working well, then your body won't be able to properly stabilize and protect itself when you lose your balance. Not a good situation.

So how do you work out fast-twitch muscle fibers? Strength training.

Now, I don't mean that you have to get a membership at Musclehead Gym. Use weights that challenge you but won't cause injury. And what if you are in your fifties, sixties, or beyond and haven't done strength training for a long time (or ever)? My good friend Debra Atkinson, MS, CSCS, one of the country's leading experts in exercise over 50, recommends starting with three simple exercises that will work out those fast-twitch muscle fibers: squat or leg press, chest press, and rows. These three basic exercises work many muscle groups and are the perfect place to start.

So now that you've begun strength training, what else can you do to slow down the aging process, protect yourself from injury, and keep your body balanced and limber as you age? One of my favorite activities: yoga.

Unless you live in a cave, I'm sure you know what yoga is: it's an ancient practice from India which involves holding or moving through different positions (called asanas) for mental and physical health. Yoga has been widely studied, and according to researchers, it can have major health benefits, both physical and mental. In a 2011 study published in the *International Journal of Yoga*, researchers wrote, "Yogic practices enhance muscular strength and body flexibility, promote and improve respiratory and cardiovascular function, promote recovery from and treatment of addiction, reduce stress, anxiety, depression, and chronic pain, improve sleep patterns, and enhance overall well-being and quality of life."[208] Sounds good to me!

A more recent meta-analysis from 2019, published in the *International Journal of Behavioral Nutrition and Physical Activity*,[209] looked at how yoga would affect older adults in good health based on multiple research sources and found "significant effects" in the group practicing yoga compared to the group that did not practice yoga, including better balance, lower-body flexibility, and lower-limb strength, as well as improvements in depression, perceived mental and physical health, sleep quality, and vitality.

Even the National Institutes of Health recommend yoga[210] for relieving stress, improving mental-health issues like anxiety and depression, improving sleep, relieving joint pain and tension headaches, helping to reduce excess body weight, helping with the symptoms of menopause, improving quality of life for people with chronic diseases, and even helping with quitting smoking. When the NIH weigh in with a thumbs-up, you can be sure yoga has gone mainstream.

Some people are already fans, but for those who think they wouldn't like or can't do yoga, rest assured that there are many different types, from gentle yoga and chair yoga to more rigorous types like power yoga and hot yoga. Most cities have yoga classes, but there are also many yoga workouts online and via apps, some for free. Just look around and you are sure to find a type of yoga that suits you. I think it's one of the very best types of exercise due to its concentration on alignment, balance, strength, flexibility, and stress management. Yoga is truly a versatile way to work out as well as relax.

Prior to my midforties, I never practiced yoga. But several years ago I began to realize that my balance and flexibility were declining, and my entire body (especially my back) was getting stiffer. So I started taking yoga classes on the Peloton app with my fave, Kristin McGee, and have never looked back. Now I try to do yoga twice a week and find that it helps immensely and in so many ways.

I recommend doing 20 to 60 minutes of yoga 2 or 3 times a week. Some people do it every day, perhaps first thing in the morning, and that's not too much. You really can't overdo yoga, but don't push yourself too far and get injured. Being aware of your limitations is a good idea with any exercise, even walking.

Yoga has really improved my life, and I believe it has helped my body age more slowly. Give it a try and see if you notice the same autojuvenative effects.

GRATITUDE

I'm not one to journal. I've tried it before, but I'm never really able to keep it up. However, many of my friends keep a daily gratitude journal. Every morning, they wake up and jot down several things they're grateful for. If you like to journal or want to try it, you might find this is a great stress-reliever, and some people find it is really therapeutic to look back and read the things they wrote to see how far they've come or to remind themselves how blessed they are.

But you don't have to journal to feel gratitude. I try to take the time every day to think about what I'm grateful for. My health. My wife. My children. My job. My employees. My parents. My brother and sister. My dogs. My friends. My patients. My community of followers. Movies from the Marvel Cinematic Universe…kidding! (Not really.) Sometimes, when I'm feeling down or have a bad day, I take time to think about what I'm grateful for. It really puts things in perspective and reduces the immediate stress and negative feelings I'm experiencing. Try it! I think you'll find it very fulfilling.

Mindfulness Moments

Mindfulness is a meditation technique, but it's also something you can do throughout the day, and it's a great way to feel centered, grounded, and present, especially if you tend to dwell on the past or stress about the future. All you have to do is stop what you're doing for a moment and focus on exactly where you are, what you are doing, what you can see, hear, feel, taste, and smell. Just be where you are, even for 10 seconds. When you go back to what you were doing before, you may see your day from a different, calmer perspective, because really, nothing is more real than the present moment. How about trying a mindfulness moment right now?

ALTRUISM

Surprisingly, one of the best ways to feel less stress yourself is to do something for someone else. Altruism is the practice of doing things solely for the benefit of others, although altruism also benefits you because it'll likely make you feel good. Whether you volunteer, donate, pay it forward (do something nice for someone in response to someone doing something nice for you), or just open the door for someone, altruistic acts are great stress-relievers.

Research has shown that performing altruistic behaviors, like volunteering for the needy, mowing an elderly neighbor's lawn, or offering up your seat on a bus or train, has positive effects on our physical and mental well-being.[211] Altruistic deeds cause the release of endorphins, which are the so-called feel-good hormones and natural painkillers and can also reduce stress, which, as we've already discussed, is one of the great agers.

Do you have a favorite charity, foundation, or religious organization (like a local church or temple) that you support? If not, and you have the means to do so, then choosing a worthy organization to support throughout your lifetime can be a very uplifting and healthy thing to do. My wife and I have two main charitable organizations that we support: our local church and dog rescues. I'm also proud to say that my skin care and supplement business, YOUN Health & Beauty, donates a portion of profits to HAVEN of Oakland County, a shelter and resource for survivors of domestic violence. There are so many people who do much more than me for others, but I've been blessed to be able to "pay it forward" in my own small way. If you have the time and/or financial ability to support a worthwhile cause, doing so can not only help others but yourself as well! It's pretty amazing how good it feels to reach out and help.

A YOUNGER FOR LIFE ATTITUDE

Am I saving the easiest for last? Maybe, but I can't end this lifestyle section without talking about attitude. It's incredible what your state of mind can do for your physical body. One study showed that sitting up straight can reduce negative and stressful feelings,[212] and a 2021 study suggested that upright postures resulted in improved mood and better brain function.[213]

Research shows that how we think is reflected in our bodies, and how we hold our bodies is reflected in how we think. For example, if you smile, it can start to make you feel happy. If you slump, you can start to feel sad. Conversely, feeling sad can cause you to slump, and feeling happy (as we all know) can trigger an automatic smile. And interestingly, studies show Botox injections can improve mood, possibly because they can prevent you from frowning.[214] I don't write this as a reason for you to get the Tox, but it's a fascinating finding!

You can use this to your advantage, when your goal is to auto-juvenate, by doing one simple thing: thinking positive. Several recent studies[215] have shown that people with good attitudes about aging have better mental health and improved quality of life as they get older.[216] One 2022 Harvard study of 14,000 adults over age 50[217] showed that the people with the highest feelings of satisfaction about aging were the most likely to feel better, live longer, have better mental health, and have better health habits than their more negative-thinking counterparts.

There's an old song from my favorite singer, Jimmy Buffett, called "Growing Older but Not Up," and this has become a sort of mantra for many people as they age. They may be getting older, but they consciously decide not to feel old mentally. Stand up straight. Notice how you move. Are you moving old or are you moving young? Are you scowling or looking around at the world with interest and a fresh perspective?

Think about some of the stereotypical differences between old and young people. Young people tend to want to learn new things, laugh often, move more, and feel cheerful, maybe because they haven't experienced as many hard lessons and don't have as much responsibility, but that doesn't mean you can't borrow from them. Some older people like to spend time with younger people to help them stay young in their minds and hearts. They maintain a positive outlook, retain their curiosity, evolve with changing times, and don't get set in their ways.

When you try new things, your neurons will create new connections and even slow down the aging process of your brain. This is the concept of neuroplasticity, and it can keep your brain working better.

Challenging? Sure, but it's certainly possible. You can continue to take pride in how you carry yourself, how you react to people around you, how you dress, how you express yourself. Nobody is telling you that you have to act old, so why do it?

These are my top ten keys for keeping a young attitude.

1. **Stand up straight.** Try to stand, sit, and move like a young person. Don't let your body convince you that you're old.
2. **Stop being so hard on yourself.** When self-doubt and negativity creep in, take a beat and remind yourself that you're doing your best.
3. **Let yourself enjoy and appreciate life.** Life can be beautiful and full of miracles. Seek out the good whenever you can.

4. **Keep learning new things.** Brush up on a language, learn an instrument, read a challenging book, go to lectures, try a new physical activity, ask questions.

5. **Keep moving (never stop moving!).** Whether it's walking, yoga, gardening, or whatever you enjoy, keep your body moving every day, and ideally use a wide variety of muscle groups.

6. **Take breaks.** You don't have to work all the time. Build time into each week to let yourself relax and have some fun.

7. **Stretch.** Our bodies get stiffer as we age. One way to combat this is yoga (please see what I wrote about this awesome practice earlier in this chapter), and another is to stretch our bodies every day. Staying limber is one of the keys to feeling younger as we age.

8. **Forgive and move on.** One of the worst things you can do to increase the stress in your life is to hold on to bitterness and grudges. There are people who will do you wrong, who have toxic personalities, who have cost you money, time, heartache, and more. Do your best to forgive and move on. Don't let them wrong you even more by giving them any airtime in your mind.

9. **Celebrate life.** Holidays, rituals, family gatherings, friends. Lean into milestones and opportunities to connect with the people you care about.

10. **Love.** Love your family. Love your friends. Love your pets. Love your life. And especially, love yourself. Love is what keeps us all going. Indulge in love often and offer it generously. This alone can keep you feeling young and vital. If you only remember one thing from this list, let it be this: we all have a limit to the amount of money and time we can give away, but love? You always have more love to give.

Stress is unavoidable, but if you take steps to ease it and commit to habits (like a healthy diet, good skin care, good sleep, and regular exercise) that help you fight it off, I can guarantee you will feel better, and chances are very good that you will also look and feel younger—maybe younger than you have in many years. Cap it off with a youthful attitude, and you're looking younger already.

PART SIX:

REGENERATE WITH NEXT-LEVEL HOLISTIC ANTI-AGING TREATMENTS

CHAPTER NINETEEN:

DIY TREATMENTS, AND WHEN TO GO IN-OFFICE

By following the components of my holistic autojuvenative program, the vast majority of readers will be able to reverse their aging to the extent that they're happy with what they see in the mirror. Factor in the following at-home DIY cosmetic treatments, and I'd estimate that 90% of you can get where you want to go without having to set foot in a doctor's office.

Twenty-five years ago most plastic surgeons had one thing to offer you: plastic surgery. But now we offer a wide array of noninvasive and minimally invasive treatments. There is *so* much that you can do to turn back the clock and get the appearance you've always wanted without going under the knife.

But before we get to the optional, in-office treatments I'll present in the next few chapters, let's start with what you can do at home. These next three DIY treatments are a must if you want to reverse your age and autojuvenate holistically.

DIY HOME TREATMENT #1:
REGULAR AT-HOME FACIALS OR CHEMICAL PEELS

There are many mild at-home facials and chemical peels that you can buy from major skin care manufacturers. These are often great for gentle exfoliation. Just make sure to stick with the major manufacturers that you know and trust (see Appendix 2 for a list of brands that I recommend), and avoid buying unknown products online, which may not be safe. Other countries may not regulate these treatments like they do in your home country, so be aware of what you buy and from where. If you're unsure, ask your dermatologist before applying anything new to your skin.

You can also do an at-home DIY facial using products from your refrigerator. Now, I know that many of my dermatological colleagues aren't big fans of using actual food on your face due to potential risk of irritation and the belief that vetted skin care products work much better (which are both completely true!), but a DIY routine, using the acids in fruit that can act as natural exfoliants, can still be a beneficial and cost-effective alternative. Fruits like oranges, pineapples, apricots, and strawberries also contain vitamin C, which can rejuvenate and fight free radicals. Just mash the fruit and add a bit of honey, then apply to your face and relax for 20 to 30 minutes. Or try this recipe.

DIY APRICOT FACIAL

½ cup dried apricots
½ cup warm water
1 tablespoon honey

Blend all the ingredients together in a blender, then apply to your face. Leave it on for 20 to 30 minutes, then wash it off gently with warm water. Your skin will feel smoother and fresher afterwards.

* * *

Here's another simple DIY facial peel that uses orange juice and contains alpha hydroxy acids (AHAs) but is gentler than an in-office chemical peel.

DIY NATURAL ORANGE PEEL

3 tablespoons of orange juice
⅓ cup plain, unsweetened yogurt

Mix the orange juice and yogurt together, then apply to your face using a small brush. Leave the mixture on for 20 minutes, until dry, then rinse off with warm water. Your skin will feel smoother and refreshed.

It's best to repeat these types of peels once or twice a week to gently exfoliate your skin. (You could do this in place of the exfoliant in your skin care routine.) Do not repeat if your skin shows signs of allergy or irritation, and remember that most commercially available options will be more effective than what you can make at home with food.

DIY HOME TREATMENT #2:
LASERS AND RED LIGHT THERAPY

Many gadgets you can buy to use at home promise to make your skin look better. The most impressive of these devices are the at-home lasers. Typically, they function like the lasers in our offices, only much less powerful. At-home lasers can improve acne, reduce unwanted hair, and even (reportedly) reduce unwanted fat (I'll believe it when I see it!). The results vary, and these devices can be very expensive, ranging from several hundred to several thousand dollars. However, this is still likely cheaper than purchasing a series of laser treatments at your local plastic surgeon's or dermatologist's office. It also offers options for those people who don't have access to a local doctor who performs these therapies.

Keep in mind that at-home lasers have significant limitations. It will likely take many more treatments to achieve the results (if you ever do) that you can get in a dermatologist's or plastic surgeon's office. At the same time, these at-home devices may still be powerful enough to create skin irritation, burns, and even scarring. So if you try one, be sure to closely follow the manufacturer's instructions and don't buy one from a company you don't know and trust.

Although I have some reservations about at-home lasers, these concerns do not extend to at-home LED red light therapy devices. Like many others in the holistic space, I'm a fan of these DIY treatments. Red light is believed to work by strengthening the mitochondria and ATP production (adenosine triphosphate, the kind of energy mitochondria produce) in our cells. The mitochondria are the powerhouses of our cells. The idea is that by supporting our mitochondria, our cells can function more efficiently, rejuvenate themselves, and repair damage.[218] Studies have found that red light therapy can indeed rejuvenate the skin. One study found that after 12 weeks, 91% of the subjects using LED light therapy reported improved skin tone, and 82% reported increased skin smoothness in the treated area.[219] In a split-face study—a randomized, controlled trial using LED light therapy on one side of the face and a sham light on the other side, with twice-a-week treatments over four weeks—researchers found a significant reduction in wrinkles, improvement in skin elasticity, and an increase in the amount of collagen and elastin fibers in the treated areas.[220]

Red light therapy devices come in many forms: handheld devices, creepy-looking masks, rectangular tabletop systems that target your face and neck, and even full-body red light beds which some people have installed in their homes and others undergo at their local holistic medspa. Feel free to check my website autojuvenation.com for my latest recommendations on red light devices.

DIY HOME TREATMENT #3: DERMAPLANING

Dermaplaning has been one of the most popular treatments in my office since I opened it fifteen years ago. During a dermaplaning session, the aesthetician uses a special blade (it looks like a scalpel) to gently scrape away unwanted fuzzy hair, dry skin, and other superficial surface irregularities to make your skin look and feel instantly smoother. This is often followed by a "lunchtime" (meaning quick, as in you can do it over your lunch hour) chemical peel to exfoliate the skin. These treatments are typically performed monthly and typically cost less than $200.

But skin care companies have created dermaplaning devices that you can use at home. Facial shavers do pretty much the same thing, too. I am seeing these treatments being demonstrated on TikTok and Instagram all the time. I joke with my wife that dermaplaning is basically a fancy way to shave your face. Shaving with a sharp razor probably accomplishes the same thing but isn't as elegant of a treatment.

What about the fears that if you shave your facial hair, it will grow back thicker? This is a complete fallacy. The reason why it may feel thicker is because the hair that grows at the base of the shaft is naturally thicker than hair as it gets longer. So it may feel thicker since it's shorter, but it's actually not.

Holistic Beauty Blacklist

There are many cosmetic treatments that I don't recommend. These include treatments that I consider too dangerous or that I've found to be ineffective. Unfortunately, doctors are performing them every day in their offices and ORs, but I would like my readers to be informed about what to avoid. That's why I've compiled my Holistic Beauty Blacklist. This list of treatments I don't recommend is being constantly updated and can be accessed by visiting my website autojuvenation.com. Here are some of the most prominent blacklisted treatments for you to be aware of. Buyer beware!

1. Fillers made with anything other than hyaluronic acid—too dangerous
2. Thread lifts—expensive and results last less than a year
3. So-called lunchtime facelifts—these are bogus procedures
4. Mesotherapy—too risky
5. Carboxytherapy—no evidence it works
6. Lip implants—hello, duck lips!
7. Face yoga (the exercises, not the relaxation part)—yoga is great, but the jury is out on whether it does anything for the face. See the next box for details.
8. Hyaluron pen—could be the quickest way to get lumpy or scarred lips

Facial Exercises and Face Yoga

Will facial exercises, or so-called face yoga, actually lift and tighten your face? There are a plethora of videos you can watch and even apps that you can download that teach you various exercises for working out your facial muscles. Many of these claim that the exercises they offer can be "as good as a facelift" or will otherwise somehow lift your face and reduce your wrinkles. Do they work? Well, the truth is a bit complicated.

There was a recent small study of 27 middle-aged women who were taught to do daily 30-minute facial exercises for twenty weeks, then surveyed on how they felt they looked. Overall, most participants were satisfied and as a whole believed they looked approximately two years younger.[221]

The conclusion from the authors of the study is that the exercises may have improved the fullness of the cheeks and lower face by increasing the size of the facial muscles. This is the current belief behind in-office facial muscle stimulators like the Lumenis TriLift and BTL's EmFace and at-home microcurrent devices like NuFace and Foreo Bear.

But is there a downside to facial exercises?

There is. If you're going for more fullness, you might want to give these exercises a try, but there isn't any evidence they do any lifting or smoothing, and you probably have to be very regular about doing these exercises over the long-term to see any effect.

Also, be aware that the more you contract the facial muscles, the more you can create dynamic wrinkles which may become deeper with time. The whole point of Botox is that it prevents the transmission of nerve impulses to muscles—Botox basically stops you from using those facial muscles that cause wrinkles. When muscles aren't flexing, the skin looks smoother and less wrinkled. When the Botox wears off, the wrinkles come back. Dynamic wrinkles get deeper with age unless the muscles stop creating them, so essentially, facial exercises are doing the opposite of Botox.

The verdict is that overall, I'm not a fan of facial muscle exercises. Although there is some early evidence to show that it might result in an increase in fullness of the face, we know that with time it can result in the wrinkles that you have getting deeper.

However, if you still want to stimulate your facial muscles, then try an at-home microcurrent device. These work by stimulating the muscles while at rest. It's possible that you can get a similar result with these microcurrent devices as you can get with facial muscle exercises, but without the risk of added wrinkles. And no, the microcurrent won't make your face jerk uncontrollably!

If you like your facial exercises and want to argue with me, or if you decide to try microcurrent devices and want to share your experience, please let me know how it goes by leaving me a comment on one of my social media posts!

WHEN WHAT YOU'VE DONE IS STILL NOT ENOUGH

As you can see, these treatments are a little beyond the basic skin care routines in this book but still easy to do at home, if you want to take the time and buy the products or devices. Still, DIY interventions can only take you so far. What do you do if you've followed all the advice in this book *and* incorporated regular at-home facials or peels, red light therapy, and dermaplaning but still aren't happy with how much you've turned back the clock? In this case, you can visit a plastic surgeon or dermatologist.

I can't tell you how many people I know who, when they were young, vowed they would never do anything so extreme as Botox or fillers, until they began to see those signs of aging on their own faces. If you're curious about minimally invasive procedures like Botox, fillers, microneedling, chemical peels, dermabrasion, fractional lasers, photofacials, fat-dissolving treatments, and teeth whitening, in these final chapters you'll find out everything you need to know, with checklists that include benefits and risks, to help you decide if any of these treatments are for you.

But please don't jump to these in-office treatments first. You know how in college there were certain classes you couldn't take until you took other classes first? Those other classes were the prerequisites, and everything in this book up until this point is a prerequisite for what I'm about to tell you. It won't do you any good to invest your time and money into more advanced anti-aging techniques and procedures if you haven't mastered the basics of an anti-aging diet, a basic healthy skin care routine, and an anti-aging lifestyle.

But if you've done all that, then welcome to the Younger for Life Protocol 2.0. While everything in the next few chapters is still non- or minimally invasive, this is where you can intervene in the aging process more aggressively. You may never want or feel like you need to do any more than what you've already done, but if you want to go further than autojuvenation—or you're just curious about what else you might be able to do—the rest of this book will tell you what you need to know.

There are many treatments you can get in a professional setting that can truly help to turn back the clock without ever going under an actual knife. Lasers, radio frequency, and other forms of technology can really make massive changes in your skin. If you want to take a step beyond these, then we're into the injectables. Although some may argue that injectable treatments aren't completely holistic (botulinum toxin is one of the most powerful in the world), if performed appropriately, they are very safe and effective options to give you the appearance you're looking for. And I'll tell you, I know *a lot* of holistic health-care experts who undergo injections of Botox and fillers. *A lot.* Some are obvious, but many are not, and some even imply that they look the way they do naturally. If you're thinking about injections, don't feel guilty for considering them. You're not alone.

Unfortunately, there are some conditions that just can't be effectively treated without surgery, and if you have these issues and they truly bother you, you may decide that plastic surgery is an option for you. We have come very far with noninvasive and minimally invasive treatments, but they cannot do everything, so if your issue is one of the following, and you would consider surgery to do something about it, then I would encourage you to seek a board-certified plastic surgeon.

1. Droopy skin of the upper eyelids
2. Excess puffiness of the lower eyelids (bags under the eyes)
3. Severe jowling
4. Severe drooping skin of the lower face and neck

Plastic surgery is the Wild West of medicine, so do your homework and be very careful whom you choose. I recommend finding a surgeon who is a member of the American Society of Plastic Surgeons (ASPS) and a member of the Aesthetic Society. You can visit their websites at www.plasticsurgery.org and www.theaestheticsociety.org.

I strongly encourage you, however, not to jump to plastic surgery without first engaging in the more holistic, less invasive treatments I'm going to cover in the next few chapters. Always start with the least invasive treatment, and only advance if it's not working to your satisfaction. It won't do you, your health, or even your appearance much good if you don't employ the more holistic strategies first to deal with the root causes of aging.

Keep in mind, as you read these final chapters, there is no one magic bullet for everyone's aging concerns. For some people, forehead wrinkles are their main issue; for others, it's a double chin; and for others, it's age spots. Instead of presenting one unified formula, I've broken down the issues and offered my advice for treating the most common aging problems individually. I'll describe what each of these treatments are, what they do, and what you can expect.

You can learn more and see updated lists of my Dr. Youn–approved treatments by going to autojuvenation.com, but here I present you with the latest information, as of this book's printing.

CHAPTER TWENTY:

AGE SPOTS, SUN SPOTS, AND LIVER SPOTS

The first category of facial-aging issues is discoloration, namely age spots, sun spots, or liver spots. These spots can range in appearance, from light to dark, flat to elevated, small to large. They have a variety of names, including freckles, sun spots, age spots, solar lentigines, and liver spots (although they have nothing to do with the liver, just may be colored the same light brown as liver).

Sun spots are typically the result of UV radiation causing the melanocytes to increase their production of melanin as an attempt to protect your skin. Small sun spots like freckles can clump together to form larger ones. In general, they do not disappear spontaneously. If you want to get rid of them, you have to actively remove them.

Certain skin creams can reduce the appearance of sun spots. Skin lighteners, such as those that contain niacinamide, kojic acid, or licorice root extract, can be effective in lightening dark spots. Unfortunately, these creams can take upward of six to eight weeks to produce a visible effect. It is best to combine these lightening creams with an exfoliating cream (such as retinol or alpha hydroxy acids) to get better penetration of the lightening cream and therefore quicker results.

Hydroquinone (mentioned previously in chapter 13) is the most powerful skin-lightening ingredient. However, hydroquinone comes with its own potential set of problems, including rebound hyperpigmentation if used for too long and then stopped cold turkey, ochronosis (a darkening of the skin that can occur in people with darker skin), and even potentially cancerous changes in the skin, as has been seen in rats.[222] For this reason, I don't recommend hydroquinone to be used for longer than six-month stints. It might be best to avoid it altogether, however, if you are okay with a slower change. Hydroquinone is banned for use in the EU.

Whether you are considering other treatments for sun spots or not, I recommend that you regularly apply safe and natural brightening creams. They can help to suppress the pigmentary blemishes that seem to pop up more and more as we get older. And don't forget to apply sunscreen every day, especially when in the sun. This is the absolute best way to prevent unwanted pigmentation.

Now let's talk about using technology in the office to reduce your dark spots.

OFFICE TREATMENT OF SUN SPOTS

The doctor's office solution to discoloration in the form of sun spots is a safe and effective procedure called intense pulsed light, or IPL.

IPL treatments can target the dark spots and destroy them with the intense pulsed light, causing the spots to turn dark and eventually fall off. IPL treatments are, in general, painless and have no downtime. They are performed every three to four weeks, and several treatments are typically required for optimal results.

There are other treatments for age spots, such as chemical peels and laser treatments, but IPL is really the best treatment in terms of safety, effectiveness, and cost. In fact, I consider IPL treatment to be one of the best bangs for your buck in cosmetic medicine. Considering the changes you can achieve, it's not that expensive compared to many other in-office treatments, especially compared to surgery. It's also about as close to natural as you can get with cosmetic treatments, as there are no toxins, acids, or other substances involved. It's just high-intensity light doing its thing to make your skin look more even. IPL treatments go by many brand names (IPL being the generic term). These include FotoFacial, Lumecca, Optima, and BBL (broadband light, not to be confused with the other BBL—Brazilian Butt Lift).

CHAPTER TWENTY-ONE:

SAGGING, SINKING, AND DROOPING

If there's one thing (besides wrinkles) that seems to bother people the most, it is the sagging, sinking, or drooping that happens to the face with age. Gravity can get the best of us, as everything from brows to eyes to mouths, chins, and necklines can begin to move downward and cave in. Our skin sags and sinks as we age. It's a fact of life. It happens to all of us, no matter how well we take care of our skin.

We can slow this process down by following the principles of reversing skin aging I give you in this book. Instead of dealing with loose skin in your forties, you may be able to delay these issues 10 or 20 years, but eventually gravity, genetics, and aging catch up, and the skin starts to sag.

Even the most beautiful men and women in the world eventually have to deal with this, including Robert Redford, Christie Brinkley, Denzel Washington, and Michelle Pfeiffer. If your favorite celebrity is in their sixties or older and doesn't exhibit at least some saggy skin, then it's not because they have superhuman genes or apply olive oil to their skin every night, it's because they've undergone interventions to tighten or lift it. In my opinion, some good possible examples of excellent surgical procedures to tighten drooping skin may include Cher, Helen Mirren, and maybe even Tom Cruise (although I have no way of knowing if any of them did actually have surgical procedures).

Most people notice sinking in the cheeks and sagging in the neck, but skin can sink, sag, or droop in our upper and lower eyelids, our lower face (creating jowls), and the rest of our body, like our breasts, arms, tummy, thighs, and buttocks.

So what can we do about it, short of surgery? The answer is, unfortunately, not a lot. True excess skin cannot be lasered away, chemically peeled away, or Botoxed away. If you have droopy upper eyelids or severe excess skin of your lower eyelids, then unfortunately surgery may be your only really effective option for treatment.

However, that's not a first option. Consider that aging is a three-dimensional process. This is something I learned a long time ago during my fellowship in Beverly Hills. Old plastic surgery was two-dimensional and based on lifting things that had fallen. Facelifts, neck lifts, cheek lifts, brow lifts, and so on. Unfortunately, if all you do is lift what's fallen, then you're not treating a person's aging in all three dimensions. And that third dimension is what we see as volume.

SUNKEN CHEEKS

Studies show that our cheeks lose volume as we get older.[223] They appear more gaunt. This volume loss is due to a combination of thinning of skin and subcutaneous fat, shrinking of bone, and even possibly shrinking of the muscles. For this reason, re-adding volume into the face is a very effective way to turn back the clock and look younger.

When I started my practice in 2004, the only effective way

we had to add volume to the face was to perform fat injections. We would liposuction fat from the tummy, hips, or thighs, purify it, and inject it into areas of the face where the volume was lost. The good thing about this procedure is that it uses your own tissue to replace the volume to your face permanently. The bad thing is that the results can be pretty unpredictable. It's also surgery. Minor, but still surgery.

Over the past ten years, we've developed many more sophisticated options for returning volume to the face using injectable fillers. Although these fillers don't last as long as fat (only one to two years on average), they are a much easier way to help the face look softer, fuller, and overall younger.

The cheeks are one of the main locations where we lose volume with age, and the judicious injection of fillers can be quite effective to fill them out and provide a subtle lift. Restylane Lyft, Restylane Contour, and Juvéderm Voluma are FDA-approved hyaluronic acid injectables that I use in my practice for the cheeks. Restylane Lyft and Contour last about one to one and a half years, whereas Juvéderm Voluma lasts one and a half to two years on average. Because they are hyaluronic acid–based fillers, they have an antidote (hyaluronidase) in case the fillers are injected improperly or you want to get rid of them. I highly recommend the cheeks be injected using a blunt-tipped cannula (a thin tube used to deliver the filler) for maximum safety.

Another option for cheek plumping, Sculptra, is also FDA-approved for filling in wrinkles of the face. Sculptra is composed of poly-L–lactic acid and takes about three treatments for optimal effect. It creates a gradual change that can last two to three years. Now, I know what you're saying. "Wait a minute! Didn't Dr. Youn say that any fillers that aren't hyaluronic acid–based are on his Holistic Beauty Blacklist?" Yes, I did. However, Sculptra is the only exception to that rule. It has been used safely and effectively since the early 2000s, and the risks, if performed conservatively and correctly, are very low.

The key to improving the appearance of the cheeks is to avoid getting them overdone! There are so many people, especially in Hollywood, who are sporting massive cheeks, like they are saving nuts for the winter. These so-called pillow faces can look odd and puppetlike. With the cheeks (and I suppose pretty much all of plastic surgery), less is usually more.

SUNKEN TEMPLES

The temples are a part of the face that is often ignored or unappreciated with aging. They tend to become more and more hollow with age, also causing the face to look gaunt. Fortunately, treatment is fairly easy.

Injectable fillers are the best ways to revolumize the temples. I prefer using Sculptra, since it tends to create a smooth and long-lasting result (two years or more), but a hyaluronic acid filler can also be used to fill in the temples, as long as it is injected very deep—below the muscle and just above the bone. It will feel like the doctor is sticking a needle into your brain when it's performed, but don't worry. Your cranium will be in the way.

Refilling the temples can really make a positive impact toward facial aging, especially for those of us (like me) whose faces become thinner with time. If you're getting filler injected into your cheeks, then don't forget your temples, too.

DOUBLE CHINS

Some of us never lose our double chin from babyhood, and others develop a double chin as they get older. If you have an area of stubborn fat below your chin and you want to get rid of it, you don't have to go under the knife anymore. The only exception is if the double chin is combined with loose skin. When the skin becomes saggy (typically after the age of 50 to 55), then removal of fat alone still leaves you with drooping skin. Make sure to keep this in mind if you are considering getting your double chin reduced. If, however, you are younger than 50 and the skin of your neck is pretty tight, then removing the fat can be a very powerful way to reshape that part of your face.

Kybella is an injectable treatment that is FDA-approved to reduce the double chin. It's composed of deoxycholic acid, which is a naturally occurring substance in the GI tract that basically dissolves fat. I often use it to dissolve away the fat of a double chin and other small areas of stubborn fat elsewhere in the body.

A Kybella treatment takes as little as 10 minutes, as the substance is injected into the fatty area using multiple needle pokes. It causes a pretty intense burning sensation afterwards that lasts for 10 to 15 minutes. The remainder of the recovery is pretty painless, although swelling will occur and can persist for four to six weeks. I usually recommend two to four treatments for optimal results, depending on how much fat is to be reduced.

Kybella can create quite impressive swelling in some, and it's expensive, around $1350 for each treatment. That said, the results are permanent. Khloe Kardashian was once a paid spokesperson for Kybella, although to my knowledge she never actually admitted to getting the procedure herself.

There are also noninvasive options to reduce double-chin fat, like CoolSculpting and SculpSure. These treatments, although not necessitating a needle like Kybella, can create a modest reduction in the double-chin fat, but the results are subtle. Be aware that CoolSculpting comes with a tiny risk of *increasing* the fat you're trying to reduce. I once did liposuction on a woman with a massive double chin that was due to a previous CoolSculpting treatment. The risk is very low (1%, maybe less), but it's something to be aware of.

LOOSE SKIN

The jawline narrows and loses volume as we age. This is one reason why the lower face and neck tends to droop with time. If a droopy neck or jawline is your concern, know that nothing is as effective as a lower facelift to raise the lower face, jowls, and neck. However, it's a lengthy operation that can create permanent scars.

If that is not for you (and many people choose not to do this, even if loose skin bothers them), you could try injectable fillers along the jawline, which can add back lost volume and sharpen the jawline, making it look lifted. The main problem here is that it takes many syringes to do this, and the results don't last more than a year or two, meaning it's a pretty costly procedure over time.

I use Juvéderm Vollure, Juvéderm Volux, and Juvéderm Voluma for this purpose, although other HA-based fillers can be used. The results are subtle, but they can be visible and significant enough. Unfortunately, it will not work as well in someone with a wider or fuller-looking face, since the widening effect may become too exaggerated. Therefore, it's best used in people who have narrow or long-shaped faces.

Heat-based treatments can be mildly effective to tighten the jawline. When the collagen of our skin is heated to a certain temperature (41–42°C), it can become damaged. When the collagen heals, it then reforms in tighter bundles, resulting in tighter, smoother, and more youthful-looking skin. In general, the more aggressive the treatment, the more impressive the result.

Mild treatments utilize radio frequency to painlessly heat the deeper skin and usually require multiple sessions, each spaced a month or so apart. These include Pellevé, ReFirme, and Exilis. They are best for people in their thirties and forties whose skin has mild looseness.

For many years I considered Ultherapy to be the gold standard for noninvasive skin tightening. It utilizes ultrasound waves that target the deeper layers of the skin to heat them up. The results are typically more impressive than noninvasive radio frequency treatments like Pellevé and ReFirme. Although it works, the treatments can be quite painful, and results can vary depending on the patient.

Radio frequency microneedling is a more recent, minimally invasive option for tightening the skin. I consider this to be the current gold standard for noninvasive skin tightening. The needles from devices like Fractora, Vivace, and (the best option, as of this writing) Morpheus8 emit radio frequency energy, causing the collagen to tighten. It can be done in an office with some topical anesthesia and requires a minor recovery. It's one of the most popular treatments in my office today, and I've had it done myself a few times. If Morpheus8 doesn't do it, then I recommend my patients either accept the loose skin or consider surgery.

Are Jade Rollers and Gua Sha FACT or CAP?

Over the past few years I have somehow become one of social media's most prominent voices combatting misinformation in cosmetic medicine. That's okay with me because I have a lot of opinions on cosmetic procedures and beauty products, and I don't like seeing people misled or cheated.

Here's an example: I have been tagged hundreds of times in the comments section of TikTok and Instagram Reels on videos showing how to use jade rollers and gua sha stones. The claims are that these beauty tools can give people a sharp, "snatched" jawline. My followers ask me, "Are these claims FACT or CAP?" (For all you non-Gen Zers, CAP is slang for "untrue" or "a lie.")

They want to know if these treatments work. Can they create permanent changes to the appearance and architecture of your face? Are they FACT or CAP?

Let's start with gua sha.

Gua sha is rooted in traditional Chinese medicine (TCM). According to TCM, your body contains rivers of energy called the chi. When the chi becomes stagnant in certain areas, a person can develop health issues. The gua sha is a specifically shaped, smooth-edged, flat stone tool that can be "scraped" over the skin to gently massage the face and improve the flow of chi.

This is what TCM believes. But what happens from a Western medicine perspective?

When a person massages their face with a gua sha stone, especially in the direction lymph is moving under the skin, it can push excess, stagnant fluid out of the face and increase circulation. This can create a temporary but sometimes visible change. So yes, it can make you look a bit more snatched afterwards.

However, it's important to realize that massaging your face with a gua sha stone won't actually create permanent changes to the architecture of your face. It's impossible for a simple massage, no matter what shape the stone is, to cause a jowl to be permanently lifted, to carve out cheekbones, to make the fat in the deep compartments of the neck magically disappear, or to shrink and tighten the skin of a sagging neckline.

Just apply a bit of logic. The concept is ludicrous. If a simple massage could move fat from one part of the body to another, then I'd spend all day massaging my love handles down to my... Oh nevermind.

And what about jade rollers?

These small devices look similar to paint rollers, but are used to massage parts of the face, most commonly the lower eyelids and jawline. Jade rollers are most common, but some people use other crystal rollers such as amethyst, rose quartz, and obsidian. There are beliefs among the New Age communities that these different crystals can convey different benefits, such as amethyst protecting the user from negative energy and rose quartz channeling love and positivity.

In practice, these crystal rollers work in much the same way as gua sha stones. They are great for gently massaging the skin, increasing circulation, and reducing swelling, especially lower eyelid puffiness. Many people will keep their crystal roller in the refrigerator and remove it just prior to using it so they get the benefit of the cool temperature and the lymphatic massage.

It all comes down to expectations. Although gua sha stones and jade rollers are great components of a skin healthy regimen and can temporarily reduce puffiness, they do not create permanent changes to your skin. Use them if you want to reduce puffiness, but don't be fooled by extraordinary claims. Those claims are CAP.

CHAPTER TWENTY-TWO:

THE EYES HAVE IT

Like many of my patients, if there is one part of my face that I'm self-conscious about age-wise, it's my eyes. I've inherited puffy under-eyes, and although I don't have bags per se, my puffy lower eyelids are the main thing I notice in photos that age me prematurely. For you it could be your upper eyelids, brows, or something else. Whatever the case, aging often shows first around the eyes. Whether you're worried about crow's feet, under-eye circles, or drooping lids, there are some solutions that don't require surgery.

CROW'S FEET

Crow's feet are wrinkles that radiate outward from the eyes, caused by years of squinting the orbicularis oculi muscles. But are they really a bad thing? These lines are associated with the Duchenne smile, which is considered to be the best example of a genuine smile of happiness. Yet even though they're associated with smiling and happiness, they're one of the earliest signs of aging that people don't like.

Because these wrinkles are caused by repetitive contraction of these muscles over many years, combined with thinning and aging of the skin, we don't have a super-effective topical or non-invasive treatment for them. Prevention is key with crow's feet. Wearing sunglasses is a great way to prevent squinting. Good sunglasses also provide protection from UV rays, which can contribute to crow's feet.

Making a definite effort not to repetitively squint in general can also help. People who have very expression-filled faces are more prone to crow's feet. Have you seen Jim Carrey lately? He's an incredible talent whose rubber face has made him millions, but this has also led to some pretty deep crow's feet. Not a bad trade-off in the end, though!

Earlier, I mentioned using red light therapy at home, and while this won't erase crow's feet, it can help to prevent and subtly smooth them. It's extremely important to use anti-aging creams that contain retinol and/or peptides or growth factors to reduce the appearance of crow's feet and slow their progression.

Certain serums also create a pretty dramatic, albeit temporary, improvement in the appearance of crow's feet. After application, these serums dry and leave a tight film behind, kind of like when you allow Elmer's glue to dry on your skin. The crow's feet will look smoother for several hours, but keep in mind that once the serum wears off, the wrinkles will return to their original state. So if your crow's feet really bother you, and you'd like something longer-lasting, you should consider in-office treatments.

The absolute best way to treat crow's feet is with Botox or other neurotoxin injections, like Dysport and the longer-lasting Daxxify. They weaken the orbicularis muscles that cause the wrinkles. They typically work within a week and last three to four months on average, slightly longer with Daxxify. Other in-office treatments are aimed at treating the symptoms (the actual wrinkles) versus the root cause (repetitive contractions of the orbicularis muscle). These treatments, like microneedling, fractional lasers, and chemical peels, can help, but botulinum-toxin injections are definitely more effective and cost-efficient for reducing and smoothing crow's feet.

DARK CIRCLES AND UNDER-EYE HOLLOWS

In order to treat dark circles under the eyes, it's important to first figure out what is causing them. The treatment will be targeted to the cause.

PIGMENTATION

If the cause is pigmentation, most commonly seen in people with darker skin to start with, the dark circles are best treated with topical skin lighteners. I recommend ones that contain ingredients such as kojic acid, licorice root extract, and niacinamide. Combine these with a retinol eye cream for stronger penetration and quicker, better results.

Realistically, as with the treatment for sun or age spots anywhere on the face, it can take months to see an improvement, and depending on how deep the pigment is, results can be pretty disappointing even after several months. Most doctors don't perform IPL treatments so close to the eyes. So in general, I recommend sticking with the brightening and lightening creams. They take a long time to work, but with time you should see improvement. And don't forget to use your sunscreen!

THIN UNDER-EYE SKIN

In other cases, dark circles are caused by thin, nearly transparent skin under the eyes that exposes the dark blood vessels underneath. The skin of the lower eyelids is the thinnest skin anywhere on the body. In some areas, it's only a handful of cells thick. The blood vessels underneath (typically veins that are dark reddish-purple in color) can cause the area to look dark. Because the root cause is thinning of the skin, the treatment is to thicken the skin to make it less transparent, thereby reducing the reddish-purple color.

You can do this topically with retinol creams which can start to thicken the skin within six to eight weeks. However, don't apply prescription-strength tretinoin to your sensitive eyelid skin! It may be too powerful and can cause severe irritation, dryness, flaking, and even visible worsening of wrinkles. Instead, look for an over-the-counter retinol eye cream for best results. Retinol eye creams are typically less potent than retinol face creams and may be more moisturizing. Don't be afraid to layer on a good moisturizing eye cream over a retinol eye cream if your skin is really dry.

PUFFINESS

Puffiness under the eyes can cause literal shadows that appear as dark circles. This is the most common cause of dark under-eye circles, and they tend to be genetic. If you have them, one of your parents probably has them, too. As I mentioned at the beginning of this chapter, my puffy lower eyelids are probably the aging feature which I dislike the most, and I believe I've inherited them from my mother. Under-eye puffiness is caused by fat from inside the eye socket that protrudes, but it can worsen if you eat a lot of salt, suffer from allergies, or have been crying.

While avoiding salt and treating allergies can help, the only permanent solution is surgery. However, you can get a dramatic (if temporary) improvement with the topical instant-tightening serums that I mentioned in the crow's feet section. These are the only noninvasive treatments that really work impressively, but the results only last a couple of hours on average, so bring the bottle with you so you can reapply it in the bathroom!

DROOPY BROWS

Many of us look grumpier as we get older. Maybe this is because those darned kids won't stay off your lawn or every car ahead of you is going five miles under the speed limit, but more likely, it's because gravity pulls the eyebrows down with age. This often combines with frown lines between our eyebrows (the *11* sign) to make us look angry.

Botox (or one of its neurotoxin competitors like Dysport and Daxxify) can be used to create a modest brow lift. In people who have droopy brows, or brows that are relatively horizontal in shape, Botox can be injected in certain areas to create an arching or lifting of the brows. Good plastic surgeons, dermatologists, and expert injectors know how this works and should be able to create this effect without much difficulty. If your eyebrows are quite arched, this may not be a good procedure for you, however. An overarched brow may give you a wicked Cruella de Vil look. I call this the Botox Brow, and I think I've spotted it on quite a few celebs.

Ultherapy is an FDA-approved treatment for lifting the eyebrows. It works by focusing ultrasound waves deep into the skin, causing the collagen to contract and therefore tighten the skin and lift the brows. The results, like all noninvasive treatments to lift the brows, are pretty subtle and vary from patient to patient. This treatment can be painful and quite expensive, costing up to thousands of dollars at some locations.

Although skin-tightening lasers and radio frequency can theoretically tighten the skin of the forehead and indirectly lift the brows, I don't find them effective. You can try it, but the change will be subtle, like Ultherapy, and therefore I'm not giving it a big stamp of approval. Try it at your own risk—the main risk being to your wallet if you don't get the results you're hoping to achieve.

CHAPTER TWENTY-THREE:

FINE LINES, FROWN LINES, AND FOREHEAD WRINKLES

Wrinkles. When it comes to facial aging, this is what most people think of...and many dread. They are a quintessential sign of aging, and I've found that people will go to great lengths to minimize them. Wrinkles usually start as fine lines you may begin to notice in your magnifying mirror, and progress to deeper lines that some people find quite distressing. Let's start with those wrinkle precursors: fine lines.

FINE LINES

Fine lines are caused by a combination of factors, including sun damage, environmental exposures, dry skin, loss of collagen and elastin, and skin–muscle interactions. If you are concerned about fine lines, the first thing to consider is taking good care of your skin. Performing the steps of renewing and protecting with good skin care is the most important factor in reducing and preventing fine lines. So make sure you stick to those steps, as explained in chapter 14. Otherwise, fine lines are typically treated by exfoliation or removal of the upper layers of the skin. This can reduce their appearance and cause the skin to look smoother.

There are, however, things you can do to decrease the look of fine lines even more. Here are some of the in-office procedures that can make a real difference.

- **Laser treatments.** The old-fashioned CO_2 laser is the most aggressive means to smooth and rejuvenate the skin. The problem with this laser is that it is ablative, meaning it literally burns off the entire upper layer of skin to accomplish this. This leaves people feeling, not surprisingly, like a burn patient, with oozing, redness, and significant pain and downtime.

  The newer CO_2 treatments are fractional, meaning instead of burning *all* the skin, only a fraction of the skin is burned. The fractional lasers are now the standard of care for aggressive treatment of lines and wrinkles. These lasers come in many different brands and names, such as Fraxel, Active FX, Deep FX, and others. The downtime also varies, with some aggressive treatments necessitating a week or more of recovery. Sometimes, depending on how aggressive the laser is, you may need a sedative or light anesthetic to tolerate the actual treatment.

  These laser treatments can be quite costly (in the thousands of dollars) as you are typically paying for the doctor's time in addition to the cost of leasing the laser. Fractional lasers typically cost a plastic surgeon upward of $100,000 to purchase, and the cost is therefore passed on to you, the patient.
- **Chemical peels.** Chemical peels are another way to effectively smooth the skin. In general, the more aggressive the peel, the longer the downtime, and the more the change. One benefit of chemical peels is that they are typi-

cally more affordable than laser treatments. The cost of the supplies needed to perform a chemical peel can be as low as $10, compared to the cost of purchasing and operating a laser. For that reason alone, chemical peels can be considered a cost-effective alternative to fractional laser treatments. Still, aggressive chemical peels can easily cost in the upper hundreds of dollars.

The most aggressive and effective chemical peel, the phenol peel, must be done under the care of a vigilant plastic surgeon or dermatologist, because the acid can create cardiac arrhythmias if overapplied. Like the old ablative CO_6 lasers, the phenol peel is gradually going out of favor, as doctors and patients are opting for less aggressive treatments to exfoliate the skin. I do not perform phenol peels, since I'm honestly afraid of them. Fatal arrhythmia while I peel the skin off your face? No thanks!

However, in my office my team and I perform quite a few moderate-depth chemical peels. My current favorite is the ZO Controlled Depth Peel, which uses a bluish-green dye to create a more even, controlled peel. Patients peel for a week afterward, but the recovery is completely painless. The only downside to the ZO Controlled Depth Peel is that you leave the office looking like the Hulk.

Many people undergo lunchtime peels with no downtime or pain. These are great treatments to modestly smooth the skin and create some mild tightening. The results are temporary, and the peels should be repeated approximately every month to maintain their results. These peels rarely cost more than a couple hundred dollars, depending on the location.

- **Microneedling.** You may have seen handheld rollers with tiny needles that, when rolled on the skin, cause microtrauma to it. These dermal rollers puncture the skin to very tiny depths, typically less than one millimeter, causing the skin to be temporarily injured. When the skin has a limited injury, the healing process causes the collagen to heal in a tighter, smoother fashion. Lasers and chemical peels also function by the same mechanism. In the case of lasers, it's via light energy; for chemical peels, it's an acid; and for microneedling, it's tiny needle punctures.

The problem with the handheld dermal rollers is that the round shape of the roller can cause the depth of the puncture to vary. Therefore, several companies now manufacture handheld automated microneedling devices. These devices can be calibrated to create a very consistent depth of micropuncture, from one to two millimeters or even deeper. This allows the doctor or aesthetician to tailor the treatment to how aggressive the patient wants. The deeper the punctures, the more trauma to the skin and the greater the changes you will see.

Microneedling is now being combined with topically applied growth factors for an even greater rejuvenating effect. The idea is that the tiny punctures created by microneedling can act as channels into the deeper skin (the dermis). Applying a growth-factor serum immediately after the microneedling treatment allows it to penetrate deep into the skin to rejuvenate it. Many practices have seen great changes to the skin by using this potent microneedling and growth-factor combination. The skin becomes smoother and tighter and has fewer fine lines.

Some doctors are combining microneedling with the application of platelet-rich plasma (PRP). In this process, the doctor or nurse draws the patient's blood, spins it down, separates the platelet-rich plasma from the red blood cells, and applies the PRP on the treated skin. The PRP is chockfull of the patient's own, all-natural growth factors.

Now, this isn't the so-called blood facial that you may have seen. Because the PRP has the red blood cells removed, it looks more like a clear yellowish serum. This is a great way to utilize the body's own regenerative properties to rejuvenate itself. It's actually part of a rapidly expanding category of anti-aging treatments called regenerative medicine.

These microneedling treatments can take about 45 minutes in a doctor's office or medical spa and can range from $200 for a superficial treatment to upward of $1,000 or more when PRP is used. Downtime is typically a day or two, unless you have a deeper treatment. It's very important to inform the doctor or aesthetician if you have a history of cold sores so that they can prescribe you an antiviral medication to prevent an eruption after the procedure. This is also true for all aggressive laser treatments and chemical peels.

There is a relatively new category of cosmetic treatments that combines hollow-tipped microneedling with neurotoxin, fillers, and more. In my office we call it the YPS (YOUN Plastic Surgery) Gold FillTox Facial. This revolutionary treatment uses gold-tipped hollow needles during microneedling. However, instead of applying growth factors over the surface of the skin, these hollow needles allow us to inject rejuvenating substances into the superficial skin to cause it to plump up and smooth out. Here are some of the actual substances we inject:

- **Hyaluronic acid fillers**, which moisturize the skin from the inside out.
- **Botox**, which helps to smooth the skin and even reduce pore size and facial sweating.
- **Growth factors**, which rejuvenate the skin as described earlier.
- **PRP.** There is no standard recipe for this treatment, as each doctor creates their own combination that they deem best for their patients. I've created the combination that we use

in the YPS Gold FillTox Facial after consulting with doctors across the country and trying it out on volunteers. Some doctors don't add the PRP, others don't use hyaluronic acid fillers, and still others add even more to it. Each practice is unique. Because of this, it goes by different names.

FROWN LINES AND FOREHEAD WRINKLES

Frown lines (the vertical *11* between your eyebrows) are caused by creating that stern expression you make when you are scolding your kids or your dog. These wrinkles are made by muscles flexing and are what we call dynamic lines. Crow's feet are also dynamic lines, caused by smiling or squinting.

Forehead wrinkles are caused by contraction of the frontalis muscles, two large muscles of the forehead that lift your eyebrows up. Every time you lift your eyebrows, you make these wrinkles just a little deeper. The best way to treat them is Botox.

Botox is basically a toxin, used in minute quantities, that is injected into the muscles that create these wrinkles. Within a week, these muscles weaken and even become paralyzed, due to the Botox blocking the nerve impulses. This causes the muscles to relax and the wrinkles to look smoother. Botox typically wears off within three or four months, after which it needs to be repeated, otherwise the wrinkles return. You can also try one of Botox's competitors like Dysport and Daxxify, mentioned previously. Results with the latter two are very similar to those with Botox, except Daxxify might last a little longer.

Now, if you're afraid of getting Botox, you're not alone. It is a powerful toxin, it's injected with a needle, and needle pokes can hurt. But believe me when I tell you that it's not that bad. I regularly get Botox injections into my *11*s, and I barely flinch when the needle goes in. There is a Botox "crunch" that you hear and feel when the needle penetrates, which is a little weird, but otherwise I don't find it to be a big deal.

GRUMPY MOUTH, MARIONETTE LINES, AND NASOLABIAL FOLDS

Next let's turn to the wrinkles around the mouth. Grumpy mouth refers to the turning down of the corners of the mouth. Some other terms for a grumpy mouth are *resting bitch face* and *permafrown*. Genetics combines with gravity to pull the corners of our mouths down as we get older. Frequent frowning may make this worse.

Admission: my grandpa had a permafrown. My dad has a permafrown. Now I—gulp—also have a permafrown. You may not have noticed it since I purposely avoid showing it when on television or on my social media, but it's there. When I'm not smiling, the corners of my mouth droop, making me look like Grumpy the Dwarf when my face is in repose. My daughter takes photos of me unawares, then shows them to me. Yes, I hate it. So what can be done with this?

Botox can help with grumpy mouth. The easiest way to turn that frown upside down (other than just smiling, which is what I try to do when I'm on camera) is to inject a tiny amount of Botox into the muscle (the depressor anguli oris) that pulls the corners down. By injecting Botox into it, you can weaken this muscle and lessen its effect. This can result in the corners of the mouth lifting up.

For those with only a minor permafrown, Botox is often all that is necessary. However, when the frown is severe, then Botox alone might not cut it. In this case, it can often take injectable fillers to lift the corners of the mouth.

The best filler for grumpy mouth really depends on many factors, including the severity of the permafrown, the thickness of the skin, and whether it's combined with marionette lines (see the next page). For thinner skin and less severe permafrown, often finer hyaluronic acid fillers such as Restylane Refyne, Restylane Silk, and Juvéderm Volbella are best. With thicker skin and more severely downturned corners of the mouth, thicker fillers with more lifting ability can be better. These include Juvéderm Ultra, Juvéderm Ultra Plus, Juvéderm Vollure, Restylane, and Restylane Defyne. The best choice of filler should be determined by your plastic surgeon, dermatologist, or expert injector.

Marionette lines are vertical grooves that extend from the corners of your mouth down to your chin. They can vary from being quite fine and superficial to deep and severe. They are often accompanied by downturned corners of the mouth. The best way to treat these lines is with an injectable filler. The choice of filler depends on the depth of the marionette line. For thinner, more superficial lines, I recommend a finer filler, whereas for deeper marionette lines, a thicker filler with more lifting capability is better.

The vertical lines of the upper and lower lip are often called *smoker's lines*, although they happen to almost all of us eventually. Smokers can definitely get them earlier than nonsmokers, and they're often deeper. When the orbicularis oris muscle contracts, it allows you to purse your lips, whistle, kiss, and suck on a straw. Doing a lot of these activities can also theoretically cause you to have more smoker's lines as well. So stop making out with your boyfriend so much or you'll get smoker's lines! Joking! (Kind of.) It could happen, but love > wrinkles.

Like all aging issues, the most effective way to treat mouth wrinkles is to deal with the root cause of these lines—the orbicularis oris muscle. Preventing it from contracting by not pursing your lips, smoking, and drinking out of straws and bottles can definitely help prevent them. However, once you have them, then the only effective way to address the root cause is with Botox.

Some practitioners inject small amounts of Botox into the orbicularis oris muscle to prevent it from contracting. Although this works theoretically, real life is more complicated. The mouth is made to move, and when its movements are limited by Botox, it can cause your mouth to look unusual and unnatural. Think about saying the word *water*. You need to purse your lips to make a *w* sound. How would you pronounce it (and how would you look) if you couldn't make the sound of a *w*? It would definitely be noticeable.

For this reason, I am not a fan of injections of Botox into the mouth unless it's done super conservatively. Mouth wrinkles are really one of those areas in which I recommend you treat the symptoms (the actual lines) over the root cause (contractions of the orbicularis muscle). The most effective treatments for these lines are therefore treatments that smooth the wrinkles by essentially sanding them down. This can be done physically by dermabrasion and microneedling, chemically by chemical peels, or via heat through fractional laser treatments.

All of these treatments, in one way or another, function to remove or damage the upper layers of skin, causing the lines to become less deep and the skin to look smoother. They all entail some type of recovery time, ranging from one to ten days, with the greater change typically occurring with longer recoveries. If mouth wrinkles are quite deep, then they may be amenable to treatment with injectable fillers. There are a few very fine fillers (Restylane Refyne, Restylane Silk, Restylane Kysse, and Juvéderm Volbella) that can be very effective at filling in deeper mouth wrinkles.

Dr. Youn's Pro Tip: Know the Risks of Facial Injections

Early in my practice, I had a beautiful young woman in her twenties come to see me for filler injections of her nasolabial folds. She disliked how these wrinkles made her look, so she drove over two hours for an appointment at my office. Just as I was about to inject her, the power went out in the building. I told her that we would have to reschedule, but she pleaded with me to go on with the treatment since she had waited so long and driven so far.

Against my better judgment, I brought her over to the

window (for more light) and injected her. Two days later, she called, concerned because her nose was completely pale and extremely painful. I had inadvertently injected the filler into her facial artery and she was suffering ischemia, or lack of blood supply, to parts of her face.

Fortunately, with some very quick and aggressive intervention, I was able to melt most of the filler away and prevent permanent damage. We were so lucky. It could have been much worse. She could literally have lost parts of her nose and upper lip.

This was a very extreme case, and the only time I've had something this serious happen after any injection in over 18 years of practice. However, it reminded me to always be super careful when injecting my patients. Here are some suggestions to keep *you* safe if you consider having injectable fillers:

1. Ask your injector to use a hyaluronic acid–based filler, like one of the Restylane or Juvéderm products. These have an antidote (hyaluronidase) which can melt away the filler if you get a complication.
2. Ask your injector to use a cannula instead of a needle if possible for more safety. Its blunt tip may lower the risk of inadvertent injection into a blood vessel. I used a needle in the patient I described above, as cannulas weren't in widespread use back then.
3. Make sure your injector is fully trained and supervised by a real plastic surgeon or dermatologist. Be careful with medspas and bargain-basement cosmetic centers, as they may not have the supervision of a doctor who has experience and knowledge to successfully treat real complications.
4. Don't get injected by the light of a window (insert image of me slapping my forehead and rolling my eyes here!).

Finally, nasolabial folds (often confused with marionette lines) are the grooves which extend from the sides of the nose to the corners of the mouth (whereas marionette lines extend from the corners of the mouth down). Nasolabial folds are often referred to as *smile lines* or *laugh lines*, although most people don't smile or laugh when they think about them.

These are wrinkles that develop due to connections between the skin and the deeper structures of the face. As the cheeks sag with age, these grooves get deeper. These are considered static wrinkles, as opposed to dynamic wrinkles, because they aren't caused by muscles and are present at all times (although they may worsen with smiling and frowning). Nasolabial folds are some of the most common cosmetic complaints that people have as they get older, right up there with frown lines and a saggy neck.

Fillers are the best treatment for nasolabial folds. Because these lines are basically deep grooves in the skin, injecting a filler can effectively push the groove out, causing the nasolabial fold to become blunted. The key to this procedure is to use enough filler to actually push the wrinkle out. Most fillers are sold by the 1-cc vial, such as Juvéderm and Restylane. One vial will only correct the shallowest of nasolabial folds. If your nasolabial folds are at all severe, you will need to consider two or even three vials for full correction. Anything less may be disappointing.

For nasolabial folds, my go-to fillers are Juvéderm Ultra, Juvéderm Vollure, Restylane, and Restylane Defyne for shallower nasolabial folds. For deeper ones, I prefer an injectable filler that is thicker with more lifting capacity, like Juvéderm Ultra Plus and Restylane Lyft.

If you decide to get fillers for nasolabial folds, I do have one caution for you: make sure your doctor uses a cannula instead of a needle if possible. There is a major blood vessel at the uppermost region of the nasolabial fold. If a filler is accidentally injected into it, you could develop major complications, ranging from scarring to loss of parts of your nose or lip to blindness. Yes, blindness. You've been forewarned.

One final tip. I've been finding that I'm injecting nasolabial folds much less than I used to. I've found (as have many other doctors) that injecting filler to plump and subtly lift the cheeks tends to improve the appearance of the nasolabial folds, often enough to prevent the patient from wanting the nasolabial folds injected, too. So, if you're considering getting your cheeks *and* your nasolabial folds injected, then start with the cheeks. You may just find that after the cheeks are enhanced, the nasolabial folds won't need anything done to them at all.

As you can see, these in-office treatments may be more effective, but again I want to remind you that much of this (including the pain, the risks, the recovery, and the cost) can be avoided if you get the basics of a healthy diet, lifestyle, and skin care program in place *first*. Only after you've already done this should you (in my humble opinion) consider these more aggressive treatments.

CHAPTER TWENTY-FOUR:

PLUMPER LIPS, WHITER TEETH

Moving beyond wrinkles, the other areas that many people feel unsatisfied about are their lips and their teeth, so let's talk about those. First, an area that has become trendy to alter, even for (or especially for) young people: thin lips.

LIP PLUMPING FOR A
YOUNGER-LOOKING MOUTH

As we age, our lips do get thinner, and youth and sensuality are closely correlated with fullness of lips. Think Angelina Jolie, Scarlett Johansson, and Brigitte Bardot (not to mention the various Kardashians and Jenners). All have (or have had) full, plump lips that many have tried to imitate. Thin lips make many people think of old ladies like Aunt May in the Spider-Man comic books. (No offense to the lovely actresses who have played this part in various movies. Marisa Tomei definitely does *not* fit this bill!)

Of all the tissues in the human body, the lips are the softest, so any treatment that enhances them must keep them feeling plush to look natural. If you have your lips done, you want to make sure they pass the Kiss Test. When someone kisses you, does it feel like kissing real lips? If it feels like they are kissing a spare tire, then you've failed the Kiss Test, and your lip enhancement was a bust. How do you make sure you pass? Be sure your injector is experienced and don't get overinjected.

The most common way to plump up the lips is by using an injectable filler. I like to describe injectable fillers as liquid skin. We can inject the filler into the borders of the lips to plump them up temporarily. The first common injection was collagen. It was rumored that Barbara Hershey may have had this injected into her lips around the time she did the movie *Beaches*, which came out in 1988, a few years after the FDA officially approved their use for cosmetic treatments.

Collagen worked well but didn't last very long. In fact, most of the time it only lasted two to four months. Lip injections are some of the most painful injections we perform, so who would want to do it every couple of months? Not me! And not most of my patients.

Today, we use hyaluronic acid. It lasts longer (six months or more in most people), and if you're unhappy with it, there is an antidote that will almost instantly dissolve it away.

If you are considering lip injections, I recommend the following HA fillers. For softer, more subtle results, consider Juvéderm Ultra, Juvéderm Volbella, Restylane Refyne, Restylane Kysse, and Restylane Silk. For more dramatic and possibly longer-lasting results, consider Juvéderm Ultra Plus, Juvéderm Vollure, Restylane Defyne, and Restylane-L.

Celebrity Fish Lips

If all you need is a good surgeon to get a great lip plump, you may be wondering why so many Hollywood stars have unnatural looking lips—and you may also be wondering how you can prevent "trout pout" or "duck lips" from happening to you.

The most frequent cause of unnatural lips is altering the natural lip proportions. The lower lip should be 50% bigger than the upper lip. For some reason, many people want their upper lip larger than their lower lip. This reversal of the natural lip proportions causes the person to look like a duck. Just say no to the quack-quack!

If injections and fillers seem a bit extreme to you, you might consider topical treatments for temporary lip plumping. Many are packaged as combination lip plumpers/lip gloss. The majority cause the lips to gently swell by using one of many irritants like peppers or cinnamon. They can cause tingling and burning, and the results can last an hour or two, after which your lips go back to normal. But maybe an hour or two is really all you want or need, like when you have a date or a class reunion. Bring the topical plumper with you for reapplication until the event is over. By the time your lips deflate back to normal, no one will be the wiser!

Remember to stick with products that ideally use natural ingredients. If you are unsure whether some of the ingredients are safe for your skin, check them against the Environmental Working Group website (ewg.org) or the Think Dirty app (free to download and use).

You can also make your own simple lip-plumping solution at home. Here is my DIY recipe.

DR. YOUN'S DIY LIP PLUMPER

Add one or two drops of organic food-grade peppermint oil (*not* peppermint essential oil—use the kind you would use in cooking) into your regular lip gloss. Put the top back on and shake it a few times, then apply the lip gloss/peppermint oil mixture to your lips.

Ideally you want to feel some minor tingling that feels cool and refreshing. If you don't feel any tingling after a few minutes, add another drop, mix, and apply again. Stop when you feel it's just right. If you apply it and your lips begin to burn, then toss it out and put in less peppermint oil next time.

WHITER TEETH, BRIGHTER SMILE

Teeth are the first thing I notice about someone. Are they white, straight, and youthful? Or are they brown or gray, crooked, and stained? As we age, our teeth get exposed to all sorts of things—coffee, tea, red wine, dark chocolate—all of these are good for you (in moderation), so I don't want you to stop enjoying them! Instead, take care of your teeth and get them looking youthful with the following tips.

The first step to a nice, white smile is to make sure your teeth are healthy. Your diet plays a big part of this, but since I've already covered that in an earlier section, let's take a look at dental care.

TOOTHPASTE

The main purpose of brushing your teeth is to remove debris and scrub away the film that can develop on your teeth throughout the course of the day (plaque) which can harden into yellow tartar over time. Toothpaste does a good job at this, generally. The most popular brands, Colgate, Crest, and others, contain a combination of fluoride to prevent cavities and sodium lauryl sulfate to create the foaming that we are all used to. Sodium lauryl sulfate is probably safe in small amounts but is known to be an irritant to the skin.

The real controversy with toothpaste, among holistic-medicine practitioners, is the fluoride. Fluoride has been added to drinking water since the mid-1940s to reduce tooth decay, especially in children. Since then, rates of tooth decay have consistently dropped over the years, causing many to credit fluoride in our water. However, as evidenced by a report in a 2016 issue of *Harvard Public Health*, the rates of tooth decay have consistently dropped in similar amounts in countries that don't have fluoride in the water.

A 2014 report designated fluoride as a neurotoxin that could be hazardous to child development,[224] and a 2017 study found fluoride exposure in pregnant women may result in their children having lower cognitive scores.[225] If fluoride is present in excessive amounts, it can lead to skeletal fluorosis, which can cause painful injuries to bones and joints, including a higher risk of fractures in older people.

Yet many groups support fluoride in water. In fact, the CDC has named fluoridation of water one of its 10 great public-health achievements of the twentieth century. However, more and more studies are being released that link fluoride to increased cancer risk,[226] acne,[227] and dental fluorosis, a condition where the tooth enamel is irreversibly damaged.[228]

So should you use fluoride-based toothpastes? I would say yes. Most dentists believe that fluoride is essential to a good toothpaste, but I believe that it may not be necessary in our water. You aren't swallowing your toothpaste (or shouldn't be!). Also remember to floss after every meal and before bedtime. I admit that I'm not great with flossing, so if you're like me, then at least try to floss before bed every night.

CROOKED TEETH

If you had crooked teeth as a kid, you may have had braces and problem solved. But some adults never got their teeth straightened or developed crooked teeth later in life. If this has been the case for you, it's not too late to get braces, and I recommend

Invisalign. These are molds specialized for your teeth to gradually make them straighter. Every two weeks the dentist gives you a new mold to straighten the teeth even more. The entire process takes about a year or less and can be very effective but also expensive.

DISCOLORATIONS OR IRREGULARITIES

If you have a discolored tooth or a tooth with a chip or irregular shape, or if your teeth are generally more discolored or irregular than you would like, there are ways your dentist or dental surgeon can even out your smile:

- **Bonding.** Bonding is a simple process that is cheaper and easier (but less permanent) than getting veneers. It's ideal for treating a few problematic teeth. A resin is applied to the tooth and polished to the ideal shape. If a tooth is crooked or shaped poorly, then it can be applied to the entire tooth to make it look completely different. Bonding lasts about three to five years.
- **Porcelain veneers.** If you want the look of perfect, straight, bright white teeth, you could consider porcelain veneers. Most celebrities who have perfect teeth have them. They require two appointments. At the first, the teeth are filed down on the front surface and a mold is made. The dentist will put bonding onto your teeth so they look good until the next appointment.

  When you return, the permanent porcelain veneers are placed, typically on the front of the tooth for cosmetic improvement. You can save money by only having veneers applied to the front parts that are visible. Veneers can last 20 years, and they don't typically stain. However, they are very expensive.

  When choosing a veneer color, I recommend choosing

one that is slightly off-white. When veneers are too white, they look unnatural—like Chiclet teeth. If you see your favorite celeb with impossibly white, perfectly straight teeth, then they are almost positively veneers, caps, or crowns.

CROWNS AND CAPS

If you want to go even further toward dental perfection, you could get crowns or caps, although this is quite an aggressive procedure. Basically, 60 to 70% of the tooth is filed away, and porcelain or metal or a combination of the two is applied over the small nubbin of the tooth. Unlike a veneer, which usually only covers the front of the tooth, a crown encases the entire tooth. However, you will lose two to four times the amount of your natural tooth compared to veneers.

The best reason to consider a crown or a cap is when there has been significant damage to a tooth, such as by grinding or a root canal. Because it is so invasive, I don't recommend crowns unless absolutely necessary.

By the way, I once went to a dentist and asked about a whitening special they had. It was only $99 for a laser treatment followed by a little hydrogen peroxide pen that I would use to apply the peroxide onto my teeth at home. The dentist took a look at my teeth and told me that while the laser and peroxide would help, I should really consider caps on my teeth.

At the time, I didn't know what caps were, so I listened to her try to sell these to me. Later on, when I found out that they file down your original teeth, I was horrified. My parents spent thousands of dollars on my teeth in the form of braces as a kid, followed by major orthognathic surgery (which I covered in my first book, *In Stitches*). To think a dentist would recommend that I trash my teeth to give me a fake new smile really made me upset. Just my opinion.

WHITENING

Discolored or yellowing teeth may be one of the most common complaints to dentists because we all see those movie stars with their (fake) white smiles and think that's how our teeth should look. Now that you know most perfect-smile celebs have veneers, caps, or crowns, you might be satisfied with simpler whitening procedures that do make a difference, even if they won't give you star-power teeth. These range from at-home self-care to in-office treatments that will make more of a difference but also cost more. Here are your basic whitening options:

- **Toothpaste.** Some of the most effective whitening toothpastes are the ones like Crest 3D and Colgate Optic White. These toothpastes work by being abrasive and scrubbing stains off teeth. There are some concerns about using these more aggressive toothpastes for long periods of time, in that they can adversely affect the enamel of the teeth.[229] Therefore, I don't recommend these toothpastes for long-term use.

  Activated-charcoal toothpastes are also very popular. Activated charcoal is very effective at absorbing and trapping unwanted substances and expelling them from the body. This is why it is a treatment for drug overdose, but it's also effective for removing surface stains from your teeth. The main problem with these toothpastes is the same as for the other aggressive whitening toothpastes: their abrasiveness. Although less abrasive than some whitening toothpastes, overuse could potentially result in permanent damage to your enamel.

  For these reasons, the whitening toothpastes I mainly recommend are the activated-charcoal ones, but only for very limited periods of time—just long enough to see some good whitening of your teeth. First discuss with your dentist whether you have thin tooth enamel, in which case no abrasiveness is appropriate for you. Then, make sure that the activated-charcoal toothpaste you use isn't overly gritty.

- **Whitening strips.** Whitening strips, such as Crest 3D Whitestrips, can work as well as at-home whitening trays or even in-office whitening. Wear them for 30 minutes once a day, and you will likely see results within a week. This product uses hydrogen peroxide, which can be irritating to your teeth and gums and lead to tooth sensitivity, which is often temporary but can last many months for some people. These strips typically only cover the front six teeth.
- **At-home whitening trays.** You can get whitening trays from your dentist to use at home. The dentist makes a mold of your teeth, sends the mold to a special lab, and whitening trays are made to match your teeth. You put the hydrogen peroxide solution into the trays and wear them overnight for approximately two weeks. The main problem I've had with these at-home whitening trays is gum irritation from the hydrogen peroxide.

  There is another good option, however, and that's activated charcoal applied to your teeth, not brushed into them. Primal Life Organics has developed a charcoal tooth gel that can be placed into whitening trays for a safer, less irritating way to whiten your teeth. It's a very interesting concept! Combine this with LED light therapy consisting of blue and red light for a greater change to your teeth. Studies have been done that combine the LED therapy with hydrogen peroxide, with significant improvement in the color of the teeth.[230]
- **In-office whitening.** There are many in-office treatments to make your teeth whiter, from LED laser treatments combined with trays to Zoom Teeth Whitening to the placement of veneers or crowns that are beyond the scope of this book. If you'd like whiter teeth, you can likely

get significant whitening and improvement following the steps I've outlined above.

Americans spent an estimated $30 billion on cosmetic dentistry in 2022. In my opinion, the key to a beautiful, ageless smile is maintaining oral and dental health as naturally as possible. Hopefully, the tips I've suggested in this chapter will help you save money yet leave you with straight and beautifully white teeth.

CHAPTER TWENTY-FIVE:

HAIR RAISING

About a year ago I went on a long-awaited trip to Puerto Rico with my wife and children. We were on a tour of the rainforest, and as I was about to jump off a cliff into the water below, my wife exclaimed, "You're losing the hair on the top of your head!"

"What??? Nooooooo!!!" (Insert horrified look here.)

I've always prided myself on my thick mane of black hair. I may have a gummy smile, a burgeoning permafrown, and early under-eye puffiness, but doggonit, at least I have a ton of dark hair. Well, not anymore. So I freaked out. It was about the closest thing I've had to a midlife crisis. That evening I had my daughter take a photo of the top of my head and it confirmed my fears: the scalp at the top of my head was showing.

So like the gold medalist snowboarder Chloe Kim about to destroy a half-pipe at the Olympics, I sprang into action. I undertook several of the steps that I'm going to share with you in this chapter. Now, before I tell you exactly what I did, let me explain all the various holistic options to treat thinning hair in case you, like me, are dealing with it right now.

Society has harsh expectations for both women and men when it comes to hair. Women are expected to have thick, shiny, luxurious hair on their heads, but not anywhere else. Men are expected to keep their hair, too, and chest hair may or may not be considered attractive. But back hair? No way. A few male celebs have convinced us that bald heads *can* be sexy, but you also need Dwayne Johnson's body, Bruce Willis's charisma, or Andre Agassi's talent to pull it off. Crap, I don't have any of those. It's not fair, and it's enough to make a hair-challenged person despair.

Fortunately, these conventional attitudes are finally starting to change, and people are tending to embrace their so-called imperfections with more confidence, with women going proudly gray and men going proudly bald (gray hair has always been considered distinguished in men—also unfair!). However, when women begin to experience thinning hair and even balding—more common than you might think—people still tend to look askance, and that can erode confidence.

I hope that someday appearance becomes less important to society at large, but until that happens, I hope I can help you with any hairy issues that may currently be undermining *your own* self-confidence, especially as they apply to aging.

Here's a fun fact. Hair changes when hormones change. Did you notice your hair got thicker, darker, or turned curlier with puberty or pregnancy, then got straighter and thinner with menopause? Women as well as men can experience hair loss with age, and just about everybody gets gray hair at some point. (Although a visit to the salon can take care of that, you'll probably need to go more and more often to maintain the effect.)

Hair, like skin, can improve and age more slowly with treatments both internal and external. Let's look at each.

CURING HAIR LOSS FROM THE INSIDE OUT

Stress can actually cause hair loss through a temporary condition called telogen effluvium.[231] Working on your stress can reverse this condition. Check out chapter 18 for tips on how to reduce stress, including some of my favorites like yoga and meditation.

What you eat can also influence your hair quality, specifically your protein intake. Some people who change their diets and stop eating enough protein (i.e., those who switch to a plant-based diet and are not mindful of their nutrient intake) may notice that their hair is thinning. This, too, is reversible: start eating more protein, including plant-based sources like nuts and beans, and your hair should begin to grow back.

Other nutrient deficiencies can influence hair quality and hair loss. If you're low on biotin, zinc, selenium, iron, and vitamins C, D, B_{16}, and others, your luxurious mane might not be as luxurious as it could be. Supplementing with these and other nutrients can help to thicken hair that is getting thinner. If you have thyroid disease, make sure to talk with your doctor before taking a biotin supplement, as it may alter the results of certain thyroid blood tests.

So how do you know which supplements to take in order to improve your nutrient deficiencies? You can consult with a dermatologist or a holistic medicine physician to test your nutrient levels, or you can consider taking an all-in-one supplement for thinning hair. My favorite one, and the one you may have already heard of, is Nutrafol. Nutrafol makes daily nutritional supplements specifically formulated for men and women at all stages of life. Their studies show significant hair growth after daily supplementation for at least three months. Another popular thinning hair supplement is Viviscal, but it's not as well-known as Nutrafol.

Finasteride, otherwise known as Propecia, is a prescription pill for men that can also help with male-pattern hair loss. By decreasing the amount of the hormone dihydrotestosterone (DHT) in the body, it encourages hair growth on the scalp. It's not approved for women, and I only recommend it for men if the other, more natural options I share in this chapter are unsuccessful.

Hair loss can be caused by an autoimmune disorder called

alopecia areata. This is a condition where hair falls out in small clumps, often the size and shape of a quarter. The amount and locations of hair loss are different for everyone. Actress Jada Pinkett Smith has come forward with her struggles in dealing with this condition. Unlike the thinning hair that can come with aging and nutritional deficiencies, alopecia areata may benefit from medical treatments like corticosteroid injections. If your hair is falling out in clumps, then I strongly recommend you see a dermatologist.

CURING HAIR LOSS FROM THE OUTSIDE IN

One of the easiest treatments you can use at home for thinning hair is minoxidil (Rogaine). This topical medication is available over the counter at most drugstores and even at Walmart and Target. Initially it was only available at 5% strength for men and 2% for women, but now a 5% women's strength is also available. It definitely works. But what if you don't want to use a medication on your scalp?

There are a number of natural remedies for thinning hair, although quite possibly the only one that has some scientific support is topical rosemary oil. One study of 100 patients found that the application of rosemary oil to the scalp had similar efficacy as minoxidil for thinning hair.[232] Although neither showed a significant increase in hair at baseline and at three months of treatment, both had a similar increase in the amount of hair at six months. Minoxidil had a greater amount of scalp itching than rosemary oil, however.

So if you'd like to try a natural therapy for thinning hair, then rosemary oil is a great option. You can mix five drops of rosemary essential oil with a teaspoon of a carrier oil like coconut oil or jojoba oil, or you can mix it into your shampoo. There are also commercially available shampoos that already have rosemary oil mixed in. Do not apply rosemary essential oil directly to your scalp because it's too concentrated and may cause irritation.

Another topical product that can really help support hair growth is formulated by a company called Hair Prescriptives, founded by one of my plastic surgery mentors, Dr. Steven Ringler. Their HPx Ekakshi Oil Complex Active Hair and Scalp Serum Treatment contains Ekakshi oil, extracted from a very rare one-eyed coconut in the Philippines. It has been shown to improve hair density, texture, and volume. This product also contains a plethora of natural nutrients to increase blood flow to and oxygen levels in the scalp and to nourish the hair follicles.

There are other topical products that people have traditionally used to improve hair thinning but aren't necessarily backed up by many studies. This doesn't mean they don't work, it's just that they don't have the studies like minoxidil and rosemary oil to back them up. These include applying anti-inflammatory and antioxidant oils like coconut oil, castor oil, and lavender oil, washing your hair with rice water, and using a derma roller on your scalp.

There are also a variety of treatments and procedures that you can add to minoxidil and/or topical rosemary oil to get even better results. Let's get into them, because these treatments can make a big difference for people dealing with thinning hair.

Low-light laser therapy is a kind of "cold laser" that painlessly stimulates hair growth, and it's noninvasive. Unlike the lasers we use to zap wrinkles or remove hair, low-light lasers do not create any heat (hence the term *cold lasers*). These devices are believed to work by infusing energy and increasing circulation to the hair follicles and scalp, thereby supporting the mitochondria (the powerhouses of the cells) and causing your hair to go into a growth phase. Many studies have shown that low-light laser therapy is effective in treating thinning hair,[233] but the devices are costly, and it can take six months or longer to see results. Some of the brand names are HairMax LaserComb, Capillus, and iRestore. They can cost $300 or more.

Platelet-rich plasma (PRP) injections are a minimally invasive way to promote hair growth. The treatment starts with a technician drawing your blood and then spinning it down in a centrifuge to separate out the plasma, which is full of platelets and growth factors. This platelet-rich plasma is then injected into areas of the scalp which have thinning hair.

One meta-analysis of 776 women who underwent PRP injections for thinning hair found a significant improvement in hair density after the treatments compared to the control group.[234] Although further studies are needed to learn more about how this treatment works, we do know that it is effective. Most doctors are recommending the treatments be repeated every four to six months to maintain results. Costs can vary, but it's not cheap. You could spend $500 or more on each session of PRP injections into your scalp.

All of the treatments in this chapter help people who have thinning hair, but they're not effective to grow hair where there just isn't any. If your hair isn't thinned but is just gone, then the only real option may be surgery, specifically hair transplants. However, if you still have some wispy hair growing in those problem areas, then doing a combination of the treatments I've just described could possibly help you.

WHAT I DID FOR MY THINNING HAIR

So what did I do when I discovered the hair on top of my head was thinning?

I made the decision that I didn't want to take or apply any medication, whether over-the-counter or prescription. I wanted to treat my thinning hair naturally, but only with clinically proven methods.

First, I had to reduce my stress level. As a surgeon, I'm always under stress, worrying about my patients. It's not that they have many complications (in fact, I think my practice has an extremely low complication rate compared to some other practices I've experienced), but I'm always worried that a patient might have an issue and am constantly thinking about ways I can make sure they are doing well. So short of not operating anymore, which is not a reasonable option, I had to find other ways to reduce my stress level. Meditation and yoga it was!

Second, I started taking Nutrafol men's supplements. Although I have my own supplement for healthy skin, hair, and nails, I wanted to take one that was purely focused on growing my hair. I took four capsules each day and combined it with a scoop of my YOUN Beauty Supplemental Collagen into a hot drink most mornings. Now, I haven't mentioned taking a collagen supplement for thinning hair in this chapter because there aren't any scientific studies I know of that have assessed whether it might help, but collagen supplements are filled with amino acids, which may help with hair growth in general.

Third, I began low-light laser therapy with the iRestore Professional Hair Growth System. I wore a helmet with 282 lasers and LEDs for 25 minutes every other day. It felt a little warm on my scalp, but otherwise I couldn't feel anything.

And fourth, I started washing my hair daily with a shampoo that contained rosemary oil.

And that's it. I didn't do any other treatments. After four months, I began to notice my hair getting a little thicker. The improvement has been slow but steady, and now that it's been almost a year, I find that my hair is realistically 40% thicker than it was when I discovered the thinning.

I've recently made some changes to my previous protocol, however, which might get me even better results naturally. Now I'm washing my hair with the Nutrafol Root Purifier Scalp Shampoo and the Nutrafol Strand Defender Conditioner instead of rosemary oil shampoo. I'm also applying the Hair Prescriptives HPx Ekakshi Oil Complex Active Hair and Scalp Serum Treatment. And I'm considering PRP injections, although I haven't made the plunge, yet.

I strongly believe all of the options I've shared with you in this chapter can help most people improve their thinning hair. It helped me! For more information on the above treatments I used on my own hair, including how to get them for yourself, please visit autojuvenation.com.

We're nearing the end of the book, and I hope I've given you some interesting information about what you *could* do to take your autojuvenation to the next level. Let's wrap this up, after all this technical talk, with a reality check about what really matters in this life, when it comes to living, aging, and feeling like your life has been fulfilling and worthwhile.

CHAPTER TWENTY-SIX:

WHAT REALLY MATTERS

In this final chapter, I'd like to offer you my parting thoughts on what's really most important in life: the people you love, the relationships you make, and the effect you have on others during this short and exciting time you get to spend alive on this earth, as the unique being that you are. None of it has anything to do with what you see in the mirror, although it potentially has a lot to do with how you feel inside, both in terms of inner purpose and mental and physical health.

Outside of my profession, I do what I can to make my own life meaningful. One of my biggest passions, you might be surprised to learn, is dog rescue. My wife and I rescue senior dogs who might otherwise have a hard time getting a home, and that has been highly rewarding for both of us.

When you live with animals, you may find that you develop a new perspective on aging. Animals live in the now. As they get older, many experience a degree of physical decline, but they don't think about it. They think (as far as I can tell) about their pack, the people they count on, the joy of running around outside or having a great meal, and the satisfaction of a really good chew toy. They don't look in the mirror, or if they do, I'm pretty sure they aren't judging themselves for having a little gray around the muzzle.

A few years ago, we rescued a little 12-year-old mutt named Sammy. We nicknamed him Num Num because he had to have a handful of his teeth removed. We found Sammy in the concrete basement of a dog shelter, where he slept on a tiny bed that reeked of pee. He had a huge bladder stone, had recently recovered from being mauled by a pack of large stray dogs, and was given up by his previous owner because her boyfriend didn't like him. Sammy's life had consisted of one traumatic event after another.

We brought Sammy home, gave him a new bed, and realized that after spending so much time in hospitals, he was no longer house-trained. So we began to diaper him, changing him several times a day. For the first few months we had him, he was skittish and nervous. He didn't trust us and even tried running away.

Then one day, it was like he let out a huge sigh of relief. Sammy realized that he was home with a family who loved him. From that day on, he was a new dog. He became the happiest little guy I've ever met. He began following me everywhere, falling asleep on my lap or at my feet. Whenever we gave him food, he would smile at us, his nubbin of a tail wagging briskly. It's as if he was saying, "Thank you for my food!" He loved going for walks and rides in the car.

Less than two years after we adopted Sammy, I found a lump in his neck. He was diagnosed with thyroid cancer, and a few weeks later, after spending a night holding him while he wheezed and coughed, we decided it was time to put him to sleep. The next day Sammy died in my arms, looking into my face, as tears streamed down my cheeks. This little guy, who had been through so much trauma and hardship in his short 14 years, still looked to me like a little puppy. To this day, my one regret (as unrealistic as it is) is that I didn't find him sooner.

I admire the way dogs age, and I think there is much they can teach us. The way they live and grow older is simple and pure. I never stop learning from them. Being there as they grow older is one way I try to do a little good in this world and give something back to the animals who have given so much to the people they love over an all-too-short lifetime.

The truth is that there is nothing any of us can do to prevent the inevitable, but as we all seek to love and be loved, to do good, and to live life to its fullest, the one thing that can help us do all of that better is to be as healthy as we can so we can feel good as we age—so we can feel younger longer. This can give us the energy to pursue our passions and take care of others, until it's our time to be taken care of—a time that will, with luck and a healthy lifestyle—not happen until the very end. Remember, think health span, not life span.

When it's my time at last, I hope that after I pass through the veil, I'll find myself walking over a grassy hill and seeing a cadre of familiar furry friends (including my little Num Num) running toward me, tongues out, tails wagging, to meet their daddy. Maybe after that, I'll see the relatives and friends who preceded me, and I'll know that my time on earth was well-spent.

But all any of us can know for sure is what happens now, and that is something you largely have control over, even if it doesn't always feel like it. I urge you to make the most of your now by taking care of your body, inside and out, so you can live a fulfilling life and know that the world is a better place for having had you in it.

And aging? Well…once you've autojuvenated what you used to think was the problematic part of getting older, then all you need to do is *live* until it's your time to move on.

May you live a lot and only age a little.

★ ★ ★ ★ ★

ACKNOWLEDGMENTS

First and foremost, thank you to God for all the blessings you've given me and my family.

To Amy. You are and will always be the best thing that happened to me.

To Mom and Dad. Thank you for your unconditional love throughout my entire life.

To Lisa and Mike (and Kristen, too). Thanks for always being there. Having you as my brother and sister means the world to me.

To Jim and Py. Thank you for being my cheerleaders.

To Wendy Sherman. I can't believe it's our fourth book together! We've come a long way. Thank you for taking a chance on me when no one else would.

To John Glynn and everyone at Hanover Square Press. It's an honor and privilege to be one of your authors. Thank you for believing in me and in this book.

To Eve Adamson. I couldn't ask for a better writing partner for this book and *The Age Fix*. I appreciate you.

To Dr. Brian Smith. I couldn't have published four books without what we did together for *In Stitches*. It's still my favorite.

To all my friends, including the Greenville guys' trip crew: Tim, Chris, Bob, Andy, Doug, and Randall. Thanks for supporting my endeavors and keeping me grounded.

To my extended family, both the Youn and Kim sides. Thank you for your support over all these years. Let's get together again soon, maybe for a cousin party!

To J. J. Virgin, Summer Bock, and Karl K. Thank you for believing in me and being fantastic business mentors. I wouldn't be where I am without your guidance and advice.

To Dr. Shawn Tassone, Dr. Jerry Bailey, and Dr. Nat Kringoudis. Thanks for having my back and all the laughs. Looking forward to changing the world together!

To Dr. Jolene Brighten, Dr. Kellyann Petrucci, Dr. Elisa Song, Dr. Mary Claire Haver, and all my friends at the Mindshare Collaborative, some of whom I've mentioned in this book. There are too many of you to list, and I fear I may inadvertently leave someone out, but I appreciate and am inspired by you.

To Andrea Livingston. Thank you for your help with all the tasty recipes. Your talent as a chef is unparalleled.

To my TikTok- and Instagram-influencer buddies (you know who you are because I follow you!), even Dr. Ricky Brown and Dr. Christian Subbio. Your content inspires and encourages me. Thank you for your camaraderie as we take over health-care social media together!

To Linnea Toney and my team at Underscore Talent. Thank you for believing in me. Looking forward to the future!

To my team at YOUN Plastic Surgery. Thank you for making me look good and being the best plastic surgery team in the entire country.

To my social media followers. Thank you for joining me each day and for all your comments, likes, shares, and selfies together. It means so much that you would allow me to be a part of your lives.

To my patients, past and present. Thank you for trusting me to be your doctor. I wouldn't be who I am without you.

Finally, to Daniel and Grace. Always do your best, be respectful and kind, and love with reckless abandon. I'm so proud to be your dad.

APPENDIX 1:

THE YOUNGER FOR LIFE DIET FOOD LISTS

Each of these foods is either nourishing, cooling, firming, or healing and many achieve more than one or even all of these things. *This list combines everything from all the Younger for Life foods listed in this book. For a free printable version which you can bring to the store with you, please visit autojuvenation.com.*

FRUIT

Apples
Apricots, dried or fresh
Bananas (especially on
    the greener side, for
    resistant starch)
Blackberries
Blueberries
Cantaloupe
Cherries
Clementines
Cranberries
Gooseberries
Grapefruit
Grapes, red and green
Kiwi fruit
Mangoes
Melons
Nectarines
Oranges
Papayas

Peaches
Pears
Pineapple
Plantains
Plums
Pomegranates

Raspberries
Rhubarb
Strawberries
Tangerines
Watermelon

VEGETABLES

Artichoke
Arugula
Asparagus
Avocado
Beet greens
Beets
Broccoli
Broccoli rabe
Broccolini
Brussels sprouts
Butternut squash
Cabbage
Carrots
Cauliflower
Celery
Collard greens
Cucumber
Eggplant
Jicama
Kale
Leafy greens, the darker
  the better
Leeks
Lettuce
Mushrooms
Okra

Olives
Onions
Parsley
Peas, green
Peppers, all types, hot or
  sweet, especially red,
  orange, and yellow
  varieties
Potatoes (especially gold
  and purple)
Pumpkin
Rhubarb
Scallions
Seaweed
Shallots
Spinach
Squash, all types of
  winter and summer
Sweet potatoes
Tomato paste
Tomato sauce
Tomatoes
Turnip greens
Turnips
Zucchini

GRAINS

Barley (unless you are gluten-sensitive)

Bran cereals

Brown rice

Brown-rice-based protein powder

Bulgur wheat (unless you are gluten-sensitive)

Corn (and corn tortillas), wholegrain (ideally organic)

Grits, wholegrain

Oatmeal, rolled oats or steel cut (if you are gluten-sensitive, look for certified gluten-free varieties)

Popcorn

Quinoa

Rye (unless you are gluten-sensitive)

Whole wheat bread and other whole wheat products like pasta, tortillas, etc. (unless you are gluten-sensitive)

Wild rice

HERBS, SPICES, AND CHOCOLATE

Chocolate: dark (over 70% cocoa), raw cacao, cacao nibs

Herbs: all fresh and dried herbs, especially dried rosemary, thyme, oregano, basil, parsley, and cilantro

Spices: all spices, especially turmeric, ginger, and cinnamon

EGGS, MEAT, POULTRY, SEAFOOD (GRASS-FED, PASTURED, OR WILD-CAUGHT WHEN POSSIBLE)

Eggs and egg whites, preferably from pastured chickens

Fish, all types, especially fatty cold-water such as salmon, mackerel, and sardines, and lean fish such as cod, halibut, tilapia, and whitefish

Game meat, such as
bison, buffalo, and elk
Poultry, including
chicken, turkey, and
wild game
Red meat, lean cuts, all
types, including beef,
pork, and lamb
Shellfish, all types, such
as shrimp, oysters,
crab, and lobster

LEGUMES/PLANT PROTEIN

Almond butter
Almond milk
Almonds
Black beans
Black-eyed peas
Brazil nuts
Butter beans
Cannellini beans
Cashew butter
Cashew milk
Cashews
Chia seeds
Chickpeas (garbanzo
beans)
Edamame, ideally
organic
Flaxseed, ground
Green beans
Hazelnuts
Hummus
Kidney beans
Lentils
Lima beans
Macadamia nuts
Navy beans
Nuts and seeds
Pea-based protein
powder
Peanut butter
Peanuts
Pecans
Pinto beans
Pistachios
Pumpkin seeds
Sesame seeds
Snow peas
Soybeans, organic
Split peas
Sunflower seeds
Tofu, organic/non-
GMO
Walnuts

FATS AND OILS

Avocado oil
Grass-fed butter
Grass-fed ghee
Olive oil

FERMENTED FOODS

Kefir

Kimchi

Kombucha

Lacto-fermented pickles

Miso

Sauerkraut

Tempeh, organic

Yogurt (dairy or plant-based, such as varieties made from soy, almond, or coconut milk)

BEVERAGES

Cocoa, dark

Coffee, black, preferably organic

Matcha

Red wine in moderation

Tea, especially green, white, and Earl Grey

AGE-ACCELERATING FOOD LIST

This list of foods to avoid or limit if you want to slow down the aging process is compiled from several lists in this book.

Burnt and charred food

Candy

Cereal: cold breakfast cereals that are not primarily fiber-based

Cornmeal

Dairy products in general

Fried food: chips, fries, doughnuts, etc.

Fruit juice

Jams and jellies

Most packaged snack foods

Pasta, white

Pastries (cake, cookies, muffins, etc.)

Pretzels

Salty foods

Soda pop

Sugar in all its forms

Sugar-sweetened and artificially sweetened beverages

Sweet tea

Syrup

Tortilla chips

Vegetable and seed oils

(corn, soy, canola,
sunflower, etc.)
White bread, rolls,
bagels, tortillas,
muffins, and most
baked goods
White rice

APPENDIX 2:

DR. YOUN-APPROVED
SKIN CARE COMPANIES

Throughout this book, I mention many kinds of skin care and I encourage you to buy from a company you trust. While I have my own skin care line (see Appendix 3), there are some companies I personally know and trust. Since it can be difficult to figure out on your own which companies are good, I'm sharing some of my favorites here.

SKIN CARE BRANDS THAT ARE
DR. YOUN-APPROVED

This list is by no means exhaustive of all the skin care lines that I approve or would potentially approve, as I'm not familiar with every single brand out there. However, if you are looking for something different than what my YOUN Beauty products offer, these companies are good places to start. Please keep in mind that this list is not an endorsement of any specific product these brands may sell, as some of these companies may still have products which contain ingredients that may be potentially harmful. Please make sure to look at the ingredient lists carefully prior to purchasing any of their products.

Note: the brands listed here are listed in alphabetical order, not in order of my preference.

LARGE BRANDS:

Clarins
Depology
Dr. Brandt Skin Care
Drunk Elephant
La Roche-Posay
Murad
The Ordinary
Vichy

MEDICAL-GRADE BRANDS:

Ourself
Skinceuticals
ZO Skin Health

NATURAL AND BOUTIQUE BRANDS:

Acure
Anne Marie Skin Care
Biossance
Cocokind
Dr. Whitney Bowe Beauty
goop Beauty
Selfless by Hyram
Tata Harper

The Spa Dr. Skin Care
True Botanicals
Versed
Youth to the People

APPENDIX 3:

WHY DID I CREATE MY OWN SKIN CARE AND SUPPLEMENT LINES?

I mentioned earlier in this book that I have my own line of skin care and supplements. Here's why.

Several years ago I realized that the basic tenets I'd learned about the practice of plastic surgery were wrong. As a young surgeon, I'd been taught that the goal was to bring people to the operating room—the bigger the operation, the better. However, surgery always comes with the risk of complications, and the worst of these is death. Because of this, I realized that the goal should be the opposite. My goal should be to keep people *out* of the operating room, using plastic surgery only as a last resort.

I had to find a better way.

So how could I do that? How could I help people feel better about how they look without putting them under the knife? I dedicated myself to figuring this out.

After years of research, I've determined that the best route to true beauty and a youthful appearance lies in an integrative, holistic approach to turning back the clock—the principles of autojuvenation—not the conventional approach, which can jump to expensive and risky surgery too quickly. As you've learned in this book, the two foundations of this holistic approach are nutrition and skin care.

Although the cornerstone of good nutrition is a healthy and balanced diet, appropriate, high-quality nutritional supplements can be major supporters. Medical research suggests that the right supplements can nourish your skin from the inside out. That's why I established YOUN Beauty supplements as my collection of the highest-quality nutritional supplements to support your skin, your aging, and your overall health.

But skin health doesn't just stop at nutrition. Department stores, pharmacies, and even doctors' offices are filled with skin care products that promise to tighten this, lift that, and even be "better than Botox." Many of these products are filled with potentially harmful ingredients and additives, like parabens, phthalates, formaldehyde releasers, and excessive fragrances. Although these products may work to rejuvenate the skin, long-term they may not be the best for your health.

On the other hand, many natural skin care product lines avoid these potentially harmful ingredients. But most of them don't contain actual effective components, and they don't make your skin look younger or healthier. So you may feel you've had to choose between products that work but contain potentially harmful ingredients or products that are natural but don't make your skin look any different after using them.

So I established YOUN Beauty skin care products as the best of both worlds. They are made with the highest-quality natural and organic ingredients but contain scientifically studied components, like vitamin C, retinol, and hyaluronic acid. YOUN Beauty skin care products are great for all skin types and cruelty-free. A portion of all profits is donated to HAVEN of Oakland County, a shelter and resource for survivors of domestic violence. What a privilege it is to support such a worthy cause together.

If you'd like to try any of my YOUN Beauty products, here is a description of each and information about how to get them.

YOUN BEAUTY SKIN CARE

Green Tea Cleanser. A versatile, moisturizing cleanser that is great for all skin types, especially those with dry and mature skin. Chamomile, aloe, cucumber, and Japanese green tea extracts are used to soothe, mildly tighten, and revitalize the appearance of the skin.

Makeup-Removing Cleansing Oil. Grape seed and pumpkin oils gently remove the day's dirt, debris, and makeup without stripping the skin of its natural oils. Follow it up with the Green Tea Cleanser for a true double cleanse.

Green Tea Toner. This superior toner contains tamarind seed extract and aloe vera to moisturize, chamomile to calm, and Japanese green tea to protect the surface of the skin. It's great to prepare your skin for antioxidants and age-reversing creams.

CE Antioxidant Serum. This vitamin C serum, which includes a 20% solution of sodium ascorbyl phosphate, is a concentrated delivery system for the highest form of active, bioavailable vitamin C. When combined with the antioxidant vitamin E, the results are a synergistic formula that soothes and revitalizes the skin.

Hyaluronic + Probiotic Serum. This product nourishes and hydrates the skin for a youthful, radiant appearance. Hyaluronic acid is a naturally occurring moisturizer that helps the skin stay hydrated and improves elasticity. The added probiotics help support the healthy balance of the skin's microbiome, leaving the skin looking and feeling more nourished and less prone to irritation.

Retinol Moisturizer. The cornerstone for any youth-promoting skin care regimen. A powerful cocktail of antioxidants combines with retinol to diminish the appearance of fine lines, wrinkles, and premature aging. It's strong enough to smooth the appearance of wrinkles but gentle enough to use every day.

Retinol Eye Cream. A powerful cocktail of antioxidants

and moisturizers combines with retinol to diminish the appearance of fine lines, wrinkles, and premature aging of the eyelids. Also contains caffeine to create an immediate tightening effect.

Advanced Exfoliating Cream. This gentle but effective polish is used to exfoliate the top layer of the epidermis, bringing newer skin to the surface. This results in skin that looks and feels immediately smoother, tighter, and more refined. Use regularly to create a long-term smoothing effect.

Peptide + Bakuchiol Moisturizer. Use this combination moisturizer to help reduce the appearance of fine lines and wrinkles. The peptides help support collagen production to improve the overall texture and tone of the skin. Bakuchiol extract has been shown to have antioxidant and anti-inflammatory properties similar to retinoids.

Calming Antioxidant Moisturizer. This soothing moisturizer contains a plethora of antioxidants to counter free radicals and oxidation. Although light and comfortable, this cream can keep even the driest skin feeling supple and hydrated.

Brightening Cream. This hydrating cream is designed to lighten hyperpigmentation, fade sun spots, and restore brilliance and clarity to drab and tired skin. Hyaluronic acid hydrates the skin, while kojic acid and organic licorice root extract combine to lighten dark spots and even skin tone.

YOUN BEAUTY NUTRITIONAL SUPPLEMENTS

Supplemental Collagen. This hydrolyzed collagen powder contains a unique blend of three patented collagen peptides to not only improve the appearance of aging skin but also support bone and joint health. It has no taste and can be easily mixed into smoothies, shakes, coffee, and other foods and drinks. I recommend putting one scoop into your Autojuvenation Jump Start breakfast smoothies.

Complete Beauty Protein Powder, Chocolate and Vanilla flavors. This is a great choice to help you maintain muscle, promote fat metabolism, and support youthful skin. The easy-to-digest protein comes from non-GMO, North American–grown yellow peas, providing an excellent array of amino acids including healthy levels of the important BCAAs (branched-chain amino acids). The vanilla flavor is the perfect protein powder to use in the morning smoothies during the Autojuve-nation Jump Start.

Skin and Hair Support. This blend of vitamins and nutrients is specially formulated to support the collagen of your hair, skin, and nails. It can be used as a substitute for a daily multivitamin.

Essential Antioxidants. A unique formula containing powerful antioxidants such as resveratrol, curcumin, and quercetin, this nutritional supplement can support overall health and appearance.

Omega-3 Fatty Acids. This is an important nutritional supplement for supporting overall health and the appearance of the skin. Our Omega-3 Fatty Acids are molecularly distilled and filtered to ensure purity and to maximize the removal of metals, pesticides, PCBs, and other contaminants.

My online store at younbeauty.com contains all of these products along with my recommended sunscreens, supplements for thinning hair, and a lot more. You can also download our YOUN Beauty app at the app store on your phone for maximal ease. **As a thank-you for reading this book, please feel free to use the code YOUNGERFORLIFE for a 15% discount on your first order!**

I hope you love my YOUN Beauty skin care products and nutritional supplements. They've been truly a labor of love for me. Thank you for your support!

ABOUT THE AUTHOR

Anthony Youn, MD, FACS, is one of the most trusted and well-known plastic surgeons in the world. Recognized as a leader in his field and followed by millions, the man known as America's Holistic Plastic Surgeon® is valued for his honest approach and ability to speak to all areas of health and well-being, not just cosmetic surgery.

A native of Michigan, Dr. Youn earned his Doctor of Medicine (MD) degree from the Michigan State University College of Human Medicine, then completed his general surgery and plastic surgery training at the Michigan State University Plastic Surgery Residency Program in Grand Rapids. He then completed an advanced aesthetic surgery fellowship with a prominent plastic surgeon in Beverly Hills.

Dr. Youn is the most followed plastic surgeon on social media, with millions of active and engaged followers. He is also the host of the most popular cosmetic-medicine podcast in the country, *The Holistic Plastic Surgery Show*, which has garnered over one million total downloads and counting.

Dr. Youn's expertise has been featured on *The Rachael Ray Show*, *The Dr. Oz Show*, *The Doctors*, *Good Morning America*, *Katie*, *The Today Show*, *CBS This Morning*, *Access Hollywood*, *Daily Blast Life*, *Fox News*, *CNN*, *HLN*, *E!*, the *New York Times*, *USA Today*, *People* magazine, *In Touch*, *Life & Style*, and *Us Weekly*, just to name a few. His public television special, *The Age Fix with Dr. Anthony Youn*, was a tremendous success, airing thousands of times over five years and viewed by millions.

A national lecturer, Dr. Youn has published several scientific articles in peer-reviewed journals and has also published *In Stitches*, his critically acclaimed and award-winning memoir of becoming a doctor (Gallery Books/Simon & Schuster, 2011), *The Age Fix*, a *Wall Street Journal* and *USA Today* bestseller (Grand Central, 2016), and *Playing God* (Post Hill Press, 2019). Named a Top Plastic Surgeon by *US News and World Report*, *Harper's Bazaar*, *Newsweek*, and *Town & Country* magazines, Dr. Youn serves on the Editorial Advisory Board for *NewBeauty Magazine* and is a member of the American Society of Plastic Surgeons (ASPS) and the Aesthetic Society and is a fellow of the American College of Surgeons.

For more information, please visit dryoun.com.

REFERENCES

1   https://doi.org/10.3389/fpsyg.2018.00067

2   http://www.amazingabilities.com/amaze7a.html

3   Davids, T. W. R., Davids, R., "The Successive Bodhisats in the Times of the Previous Buddhas," *Buddhist birth stories; Jataka tales; the commentarial introduction entitled Nidana-Katha; the story of the lineage*, 1878, London: George Routledge & Sons, pp. 115–144.

4   Jain, Vijay K., *Acarya Samantabhadra's Svayambhustotra: Adoration of the Twenty-four Tirthankara*, 2015, Vikalp, Dehradun, India.

5   https://gerontology.fandom.com/wiki/Jeanne_Calment

6   Stewart, C., Sharples, A., "Aging, Skeletal Muscle, and Epigenetics," *Plastic and Reconstructive Surgery*, 150: 4S-2, Oct. 1, 2022, pp. 27s–33s.

7   https://www.bmj.com/content/bmj/339/bmj.b5262.full.pdf

8   http://science.org.au/curious/earth-environment/animals-can-live-forever#

9   https://www.afar.org/imported/AFAR_INFOAGING_GUIDE_THEORIES_OF_AGING_2016.pdf

10   https://www.science.org/doi/10.1126/science.abk0297

11   https://www.jstor.org/stable/pdf/592433.pdf

12   https://heart.bmj.com/content/99/12/882

13   https://www.afar.org/imported/AFAR_INFOAGING_GUIDE_THEORIES_OF_AGING_2016.pdf

14   https://doi.org/10.1159/000236045

15   https://www.prb.org/resources/around-the-globe-women-outlive-men/

16   Vaiserman, A., Krasnienkov, D., "Telomere Length as a Marker of Biological Age: State-of-the-art, Open Issues, and Future Perspectives," *Frontiers in Genetics*, 2021, 11:630186. doi: 10.3389/fgene.2020.630186

17   Puterman, E., Lin, J., Krauss, J., Blackburn, E. H., Epel, E. S., "Determinants of Telomere Attrition over 1 Year in Healthy Older Women: Stress and Health Behaviors Matter," *Molecular Psychiatry*, 2015, 20(4), pp. 529–535, doi: 10.1038/mp.2014.70

18   https://www.nal.usda.gov/human-nutrition-and-food-safety/food-composition/micronutrients

19   https://www.ncbi.nlm.nih.gov/pubmed/23474627

20   https://www.ncbi.nlm.nih.gov/pmc/articles/PMC7795523/

21   https://doi.org/10.1161/CIRCRESAHA.118.313996

22   https://www.frontiersin.org/articles/10.3389/fmicb.2020.01065/full

23   https://www.ncbi.nlm.nih.gov/pmc/articles/PMC3038963/

24   https://doi.org/10.1016/j.cgh.2008.02.054

25   https://www.ncbi.nlm.nih.gov/pubmed/27793451

26   https://www.ncbi.nlm.nih.gov/pubmed/27596801; https://www.ncbi.nlm.nih.gov/pmc/articles/PMC3038963/

27   https://www.ncbi.nlm.nih.gov/pubmed/23126293

28   https://fdc.nal.usda.gov/fdc-app.html#/food-details/170904/nutrients

29   https://www.fda.gov/food/agricultural-biotechnology/gmo-crops-animal-food-and-beyond#

30   https://doi.org/10.3389/fphys.2019.01607; https://doi.org/10.1017/s0954422408138732

31   https://www.ncbi.nlm.nih.gov/pmc/articles/PMC6777918/

32   https://doi.org/10.1177/10998004211016070

33   https://www.ewg.org/tapwater/state-of-american-drinking-water.php#.W0KGxbgnY2w

34   https://www.ncbi.nlm.nih.gov/books/NBK507709/

35   https://doi.org/10.3945/ajcn.113.075663; https://doi.org/10.1007/s00394-012-0340-6

36   https://doi.org/10.1093%2Fcdn%2Fnzac068.023

37   https://www.ncbi.nlm.nih.gov/pubmed/23949208/

38   https://www.ncbi.nlm.nih.gov/pmc/articles/PMC7015462/

39   https://www.ncbi.nlm.nih.gov/books/NBK216502/

40   https://www.ncbi.nlm.nih.gov/pmc/articles/PMC2846864

41   https://www.tandfonline.com/doi/full/10.1080/16070658.2016.1216359

42   https://www.ncbi.nlm.nih.gov/pubmed/17490954

43   https://www.ncbi.nlm.nih.gov/pmc/articles/PMC2846864/

44   https://www.ncbi.nlm.nih.gov/pubmed/22944875

45   https//www.consumerreports.org/media-room/press-releases/2013/12/consumer-reports-potentially-harmful-bacteria-found-on-97-percent-of-chicken-breasts-tested/

46   https://www.ncbi.nlm.nih.gov/pubmed/22944875

47   https://doi.org/10.1016/j.meatsci.2009.04.017

48   Quaesma, M. A. G., Alves, S. P., Trigo-Rodrigues, I., Pereira-Silva, R., Santos, N., Lemos, J. P. C., Barreto, A. S., Bessa, R. J. B., "Nutritional Evaluation of the Lipid Fraction of Feral Wild Boar (Sus scrofa scrofa) Meat," *Meat Science*, Oct. 2009, 83(2), pp. 195–200.

49   https://www.ewg.org/research/pcbs-farmed-salmon#.WxRm-1SAh02w

50   https://newsroom.wakehealth.edu/news-releases/2008/07/wake-forest-researchers-say-popular-fish-contains-potentially-dangerous-fatty-acid

51   https://enveurope.springeropen.com/articles/10.1186/s12302-021-00578-9

52   https://doi.org/10.3390%2Fnu12102929

53   https://www.ncbi.nlm.nih.gov/pubmed/11293471

54   https://academic.oup.com/ajcn/article/86/4/1225/4649573

55   https://www.ncbi.nlm.nih.gov/pmc/articles/PMC5075620/

56   https://www.ncbi.nlm.nih.gov/pmc/articles/PMC3614697/

57   https://www.myfooddata.com/articles/vitamin-c-foods.php

58   https://www.ncbi.nlm.nih.gov/pmc/articles/PMC5506060/

59   https://www.ncbi.nlm.nih.gov/pmc/articles/PMC4698938/

60   https://www.ncbi.nlm.nih.gov/books/NBK92763/#

61   https://pubs.acs.org/doi/abs/10.1021/jf0344385

62   https://www.ncbi.nlm.nih.gov/pmc/articles/PMC3679539/

63   https://doi.org/10.1016%2Fj.cbi.2007.05.007

64   https://www.ncbi.nlm.nih.gov/pmc/articles/PMC4665516/

65   https://www.ncbi.nlm.nih.gov/pmc/articles/PMC2874190

66   https://doi.org/10.1155%2F2021%2F9932218

67   https://www.nia.nih.gov/health/facts-about-aging-and-alcohol

68   https://lpi.oregonstate.edu/mic/dietary-factors/phytochemicals/
      resveratrol#

69   https://doi.org/10.1161/CIRCULATIONAHA.106.621854

70   https://doi.org/10.1016%2Fj.jada.2010.03.018

71   https://www.sciencedirect.com/science/article/pii/S1568163718301193

72   https://doi.org/10.1161/CIRCULATIONAHA.106.621854

73   https://www.frontiersin.org/articles/10.3389/fmed.2022.837222/full

74   https://doi.org/10.1016/j.biopha.2018.09.058

75   https://doi.org/10.1093/ajcn.82.3.675

76   2015 Dietary Guidelines Advisory Committee, *Scientific report of the
      2015 Dietary Guidelines Advisory Committee*, Office of Disease Pre-
      vention and Health Promotion, February 2015. https://health.gov/
      our-work/nutrition-physical-activity/dietary-guidelines/previous-
      dietary-guidelines/2015/advisory-report

77   Singh, G. M., et al., Global Burden of Diseases Nutrition and Chronic
      Diseases Expert Group (NutriCoDE), "Estimated Global, Regional,
      and National Disease Burdens Related to Sugar-sweetened Bever-
      age Consumption in 2010," *Circulation*, August 25, 2015, 132(8), pp.
      639–666.

78   https://doi.org/10.1093/ajcn/86.1.107

79   https://www.ncbi.nlm.nih.gov/pmc/articles/PMC4241420/

80   https://doi.org/10.1046/j.1365-2133.2001.04275.x

81   https://doi.org/10.1093/ajcn/86.4.1225

82   https://www.ncbi.nlm.nih.gov/pubmed/17556700

83   https://www.ncbi.nlm.nih.gov/pubmed/16635908

84   https://www.ncbi.nlm.nih.gov/pmc/articles/PMC3038963/; https://
      www.ncbi.nlm.nih.gov/pubmed/15520759

85   https://www.nhrmc.org/~/media/testupload/files/low-gylcemic-
      meal-planning.pdf?la=en

86   https://www.nhrmc.org/~/media/testupload/files/low-gylcemic-
      meal-planning.pdf?la=en

87   https://www.ncbi.nlm.nih.gov/pmc/articles/PMC3551118/

88   https://doi.org/10.1016/s0753-3322(02)00253-6

89   https://theconsciouslife.com/omega-3-6-9-ratio-cooking-oils.htm

90   https://www.ncbi.nlm.nih.gov/pubmed/11293471

91   https://doi.org/10.3945/an.113.003657

92   https://www.ncbi.nlm.nih.gov/pubmed/16140878

93   https://health.gov/sites/default/files/2019-09/Scientific-Report-of-
      the-2015-Dietary-Guidelines-Advisory-Committee.pdf; https://
      pubmed.ncbi.nlm.nih.gov/24222015/

94   https://doi.org/10.1016%2Fj.toxrep.2016.08.003

95   https://www.ewg.org/childrenshealth/monsanto-weedkiller-still-
      contaminates-foods-marketed-to-children

96   https://www.ers.usda.gov/data-products/food-consumption-and-
      nutrient-intakes/

97   https://www.ncbi.nlm.nih.gov/pmc/articles/PMC3687363/

98   https://www.ncbi.nlm.nih.gov/pubmed/19496976

99   https://www.ncbi.nlm.nih.gov/pubmed/10088212

100   https:///www.ncbi.nlm.nih.gov/pubmed/12433724

101   https://www.nature.com/articles/ijo201551

102 https://www.ncbi.nlm.nih.gov/pmc/articles/PMC4229419/

103 https://www.ncbi.nlm.nih.gov/pmc/articles/PMC2892765

104 https://doi.org/10.1016/j.cell.2022.07.016

105 https://www.ncbi.nlm.nih.gov/pmc/articles/PMC2769029

106 https://www.ncbi.nlm.nih.gov/pmc/articles/PMC3704564/

107 https://www.ncbi.nlm.nih.gov/pubmed/18992957

108 https://www.ncbi.nlm.nih.gov/pubmed/18992957

109 https://www.ncbi.nlm.nih.gov/pubmed/17760202

110 https://www.cell.com/cell-metabolism/fulltext/S1550-4131(18)30130-X

111 https://www.ncbi.nlm.nih.gov/pmc/articles/PMC6036773/#ref14

112 https://doi.org/10.1016/j.arr.2016.08.005; https://www.ncbi.nlm.nih.gov/
pubmed/21402069; https://www.ncbi.nlm.nih.gov/pubmed/19590001;
https://www.ncbi.nlm.nih.gov/pubmed/24280167

113 https://doi.org/10.3390%2Fnu11122923

114 https://www.nejm.org/doi/full/10.1056/NEJMoa2114833

115 https://www.ncbi.nlm.nih.gov/pmc/articles/PMC7021351/

116 https://www.ncbi.nlm.nih.gov/pmc/articles/PMC3106288/

117 https://www.ncbi.nlm.nih.gov/pmc/articles/PMC8924556/

118 Ashford T. P., Porter K. R., "Cytoplasmic components in hepatic
cell lysosomes," *J. Cell Biol.*, 1962, 12, 198–202. 10.1083/jcb.12.1.198

119 https://www.sciencedirect.com/science/article/pii/S0092867407016856

120 https://www.ncbi.nlm.nih.gov/pmc/articles/PMC2670622/

121 https://www.ncbi.nlm.nih.gov/pubmed/15637215

122 https://www.researchgate.net/publication/5349201_

123 https://www.ncbi.nlm.nih.gov/pubmed/10702175; https://www.ncbi.
nlm.nih.gov/pubmed/18086246

124 https://www.ncbi.nlm.nih.gov/pubmed/15675947; https://www.ncbi.
nlm.nih.gov/pubmed/18494887

125 https://doi.org/10.1093/ajcn/86.4.1225

126 https://www.ncbi.nlm.nih.gov/books/NBK532266/

127 https://www.ncbi.nlm.nih.gov/pubmed/12858333; https://www.ncbi.nlm.nih.gov/pubmed/16039116; https://www.ncbi.nlm.nih.gov/pubmed/15664501

128 https://www.ncbi.nlm.nih.gov/pubmed/2997282

129 Makrantonaki, E., Zouboulis, C. C., "Skin Alterations and Diseases in Advanced Age," *Drug Discovery Today: Disease Mechanisms*, 2008, 5: e153–162.

130 https://www.ncbi.nlm.nih.gov/pubmed/20882333

131 https://www.ncbi.nlm.nih.gov/pubmed/20354654

132 https://www.ncbi.nlm.nih.gov/pubmed/26488693

133 https://www.ncbi.nlm.nih.gov/pmc/articles/PMC4065280/

134 https://www.ncbi.nlm.nih.gov/pubmed/16029678/

135 https://www.ncbi.nlm.nih.gov/pubmed/11351267/

136 https://www.ncbi.nlm.nih.gov/pubmed/28914450

137 https://pubs.acs.org/doi/abs/10.1021/acs.jafc.6b05679

138 https://doi.org/10.3390%2Fnu11102494

139 https://doi.org/10.1016/j.heliyon.2023.e14961

140 http://dx.doi.org/10.1159/000355523

141 https://doi.org/10.1089%2Fjmf.2015.0022

142 https://doi.org/10.1111/ijd.15518

143 https://www.jaad.org/article/S0190-9622(03)00781-3/abstract

144 https://www.ncbi.nlm.nih.gov/pubmed/19337100

145 https://www.ewg.org/news/testimony-official-correspondence/cdc-americans-carry-body-burden-toxic-sunscreen-chemical#.WoBxKudG02w

146 Calafat, A. M., Wong, L. Y., Ye, X., Reidy, J. A., Needham, L. L., "Concentrations of the Sunscreen Agent Benzophenone-3 in Residents of the United States: National Health and Nutrition Examination Survey 2003–2004," *Environmental Health Perspectives*, July 2008, 116(7), pp. 893–897, doi: 10.1289/ehp.11269

147 Bryden, A. M., Moseley, H., Ibbotson, S. H., Chowdhury, M. M., Beck, M. H., Bourke, J., et al., "Photopatch Testing of 1155 Patients:

Results of the UK Multicentre Photopatch Study Group," *British Journal of Dermatology*, 2006, 155(4), pp. 737–747.

148   Hanson, K. M., Gratton, E., Bardeen, C. J., "Sunscreen Enhancement of UV-induced Reactive Oxygen Species in the Skin," *Free Radical Biology and Medicine*, 2006, 41(8), pp. 1205–1212; Serpone, N., Salinaro, A., Emeline, A. V., Horikoshi, S., Hidaka, H., Zhao J., "An In-Vitro Systematic Spectroscopic Examination of the Photostabilities of a Random Set of Commercial Sunscreen Lotions and Their Chemical UVB/UVA Active Agents," *Photochemical & Photobiological Sciences*, 2002, 1(12), pp. 970–981.

149   Van Liempd, S. M., Kool, J., Meerman, J. H., Irth, H., Vermeulen, N. P., "Metabolic Profiling of Endocrine-Disrupting Compounds by On-line Cytochrome p450 Bioreaction Coupled to On-line Receptor Affinity Screening," *Chemical Research in Toxicology*, 2007, 20(12), pp. 1825–1832; Schlumpf, M., Schmid, P., Durrer, S., Conscience, M., Maerkel, K., Henseler, M., et al., "Endocrine Activity and Developmental Toxicity of Cosmetic UV Filters—An Update," *Toxicology*, 2004, 205(1-2), pp. 113–122.

150   Ma, R., Cotton, B., Lichtensteiger, W., Schlumpf, M., "UV Filters with Antagonistic Action at Androgen Receptors in the MDA-kb2 Cell Transcriptional-activation Assay," *Toxicological Sciences*, 2003, 74(1), pp. 43–50.

151   https://www.ewg.org/research/cdc-americans-carry-body-burden-toxic-sunscreen-chemical#.W0uLobgnY2w

152   Klammer, H., Schlecht, C., Wuttke, W., Schmutzler, C., Gotthardt, I., Köhrle, J., et al., "Effects of a 5-day Treatment with the UV-filter Octyl-methoxycinnamate (OMC) on the Function of the Hypothalamo-pituitary–thyroid Function in Rats," *Toxicology*, 2007, 238(2-3), pp. 192–199; Wang, J., Pan, L., Wu, S., Lu, L., Xu, Y., Zhu, Y., et al., "Recent Advances on Endocrine Disrupting Effects of UV Filters," *International Journal of Environmental Research and Public Health*, 2016, 13(8), p. 782.

153   Danovaro, R., Bongiorni, L., Corinaldesi, C., Giovannelli, D., Damiani, E., Astolfi, P., et al., "Sunscreens Cause Coral Bleaching by Promoting Viral Infections," *Environmental Health Perspectives*, 2008, 116(4), pp. 441–447.

154   https://www.ncbi.nlm.nih.gov/pmc/articles/PMC3760934/

155   https://www.ncbi.nlm.nih.gov/pmc/articles/PMC2699641/

156 Dhaliwal, S., Rybak, I., Ellis, S. R., et al., "Prospective, Randomized, Double-blind Assessment of Topical Bakuchiol and Retinol for Facial Photoageing," *British Journal of Dermatology*, 2019, 180(2), pp. 289–296, doi:10.1111/bjd.16918

157 https://doi.org/10.1080/10408440701638970

158 https://www.e-ijd.org/article.asp?issn=0019-5154;year=2013;volume=58;issue=2;spage=157;epage=157;aulast=Monteiro

159 https://ehp.niehs.nih.gov/doi/full/10.1289/ehp.1409200; https://pubmed.ncbi.nlm.nih.gov/17306434/

160 https://pubmed.ncbi.nlm.nih.gov/28692609/

161 https://www.ncbi.nlm.nih.gov/pmc/articles/PMC2866685/

162 https://www.fda.gov/cosmetics/productsingredients/ingredients/ucm109655.htm

163 https://monographs.iarc.who.int/wp-content/uploads/2018/09/ClassificationsAlphaOrder.pdf

164 Ito, N., Fukushima, S., Tsuda, H., "Carcinogenicity and Modification of the Carcinogenic Response by BHA, BHT, and Other Antioxidants," *Critical Reviews in Toxicology*, 1985, 15(2), pp. 109–150.

165 Barrett, J. R., "Phthalates and Baby Boys: Potential Disruption of Human Genital Development," *Environmental Health Perspectives*, Aug. 2005, 113(8), A542.

166 https://ntp.niehs.nih.gov/ntp/roc/content/profiles/diethylhexylphthalate.pdf

167 https://www.frontiersin.org/articles/10.3389/fenvs.2014.00036/full

168 https://www.ncbi.nlm.nih.gov/pmc/articles/PMC2794303/

169 https://www.ncbi.nlm.nih.gov/pmc/articles/PMC2699641/

170 http://news.berkeley.edu/2017/04/05/deep-sleep-aging/

171 https://www.consumerreports.org/sleep/why-americans-cant-sleep/

172 https://www.ncbi.nlm.nih.gov/pmc/articles/PMC5148237/

173 https://www.ncbi.nlm.nih.gov/pubmed/25266053

174 https://journals.lww.com/md-journal/Fulltext/2021/05210/Exercise_interventions_for_older_people_at_risk.52.aspx

175 https://www.ncbi.nlm.nih.gov/pmc/articles/PMC3754812/

176 https://doi.org/10.1111/sdi.13020

177 https://www.ncbi.nlm.nih.gov/pmc/articles/PMC5928534/

178 https://www.ncbi.nlm.nih.gov/pmc/articles/PMC5830901/

179 https://www.ncbi.nlm.nih.gov/pmc/articles/PMC4341978/

180 https://www.ncbi.nlm.nih.gov/pubmed/19645971

181 https://www.sleepfoundation.org/bedroom-environment#

182 http://www.pnas.org/content/112/4/1232

183 https://doi.org/10.1371%2Fjournal.pone.0110825; https://www.ncbi.
     nlm.nih.gov/pubmed/25758729

184 https://www.ncbi.nlm.nih.gov/pmc/articles/PMC3356284/; https://
     link.springer.com/article/10.1007/s11818-008-0353-9

185 https://www.ncbi.nlm.nih.gov/pubmed/19584739

186 https://www.ncbi.nlm.nih.gov/pubmed/26731279

187 https://www.ncbi.nlm.nih.gov/pubmed/20347389

188 https://www.ncbi.nlm.nih.gov/pmc/articles/PMC4345801/

189 https://www.ncbi.nlm.nih.gov/pubmed/16807875

190 https://www.ncbi.nlm.nih.gov/pubmed/17211115

191 https://www.ncbi.nlm.nih.gov/pubmed/22291721

192 https://www.ncbi.nlm.nih.gov/pubmed/15650465

193 https://www.ncbi.nlm.nih.gov/pmc/articles/PMC4378297/

194 https://www.ncbi.nlm.nih.gov/pubmed/15574496

195 https://www.ncbi.nlm.nih.gov/pubmed/19735238; https://www.
     sciencedirect.com/science/article/pii/S0889159113001736; https://
     www.ncbi.nlm.nih.gov/pmc/articles/PMC3057175/

196 https://doi.org/10.1016/j.ctcp.2013.10.001

197 https://www.ncbi.nlm.nih.gov/pmc/articles/PMC5783379/; https://
     www.frontiersin.org/articles/10.3389/fpsyg.2021.724126/full; https://
     doi.org/10.1001/jamainternmed.2013.13018

198 https://www.ncbi.nlm.nih.gov/pmc/articles/PMC5447722/

199 https://doi.org/10.3233/jad-142766; https://doi.org/10.1111/nyas.12348

200 https://www.ncbi.nlm.nih.gov/pmc/articles/PMC3057175/

201 http://dx.doi.org/10.21276/SSR-IIJLS.2020.6.5.2

202 https://www.ncbi.nlm.nih.gov/pmc/articles/PMC5455070/

203 https://journals.lww.com/jbisrir/fulltext/2019/09000/effectiveness_
of_diaphragmatic_breathing_for.6.aspx

204 http://dx.doi.org/10.1007/s10072-016-2790-8

205 https://doi.org/10.1016/j.invent.2019.100293

206 https://pubmed.ncbi.nlm.nih.gov/28093824/

207 Parkkari, J., Kannus, P., Palvanen, M., Natri, A., Vainio, J., Aho, H.,
Vuori, I., Järvinen, M., "Majority of Hip Fractures Occur as a Re-
sult of a Fall and Impact on the Greater Trochanter of the Femur: A
Prospective Controlled Hip Fracture Study with 206 Consecutive
Patients," *Calcified Tissue International*, 1999, 65, pp. 183–187.

208 Woodyard C., "Exploring the therapeutic effects of yoga and its ability to increase quality of life," *International Journal of Yoga*, Jul. 2011,
4(2), pp. 49–54.

209 https://ijbnpa.biomedcentral.com/articles/10.1186/s12966-019-0789-2

210 https://www.nccih.nih.gov/health/yoga-what-you-need-to-know

211 Schwartz, C. E., Keyl, P. M., Marcum, J. P., Bode, R., "Helping Others Shows Differential Benefits on Health and Well-being for Male
and Female Teens," *Journal of Happiness Studies*, 2009, 10(4), pp. 431–
448; Schwartz, C., Meisenhelder, J. B., Ma, Y., Reed, G., "Altruistic
Social Interest Behaviors Are Associated with Better Mental Health,"
Psychosomatic Medicine, 2003, 65(5), pp. 778–785.

212 https://doi.org/10.1037/hea0000146

213 https://doi.org/10.1016/j.actpsy.2021.103353

214 https://www.ncbi.nlm.nih.gov/pmc/articles/PMC5563881/

215 https://www.ncbi.nlm.nih.gov/pmc/articles/PMC8156287/; https://doi.
org/10.1093/geront/gnu041; https://academic.oup.com/gerontologist/
article/57/suppl_2/S187/3913342

216 https://hqlo.biomedcentral.com/articles/10.1186/1477-7525-11-146

217 https://jamanetwork.com/journals/jamanetworkopen/fullarticle/
2788853

218 https://www.healthline.com/health/red-light-therapy#how-does-
it-work?

219   Baez, F., Reilly, L. R., "The Use of Light-emitting Diode Therapy in the Treatment of Photoaged Skin," *Journal of Cosmetic Dermatology*, Sept. 2007, 6(3), pp. 189–194.

220   Lee, S. Y., Park, K. H., Choi, J. W., et al., "A Prospective, Randomized, Placebo-controlled, Double-blinded, and Split-face Clinical Study on LED Phototherapy for Skin Rejuvenation," *Journal of Photochemistry and Photobiology*, 2007, July 27, (Epub 2007, May 1), 88(1), pp. 51–67.

221   https://www.ncbi.nlm.nih.gov/pmc/articles/PMC5885810/

222   https://www.nature.com/articles/bjc1998492

223   https://academic.oup.com/asj/article/24/6/514/227389

224   https://www.ncbi.nlm.nih.gov/pmc/articles/PMC4418502/

225   https://ehp.niehs.nih.gov/doi/10.1289/ehp655

226   https://www.ncbi.nlm.nih.gov/pmc/articles/PMC3889024/

227   Saunders, M. A., Jr., "Fluoride Toothpastes: A Cause of Acne-like Eruptions," Letters to the Editor, *Archives of Dermatology*, June 1975, 111(6), p. 793, doi:10.1001/archderm.1975.01630180121023

228   https://doi.org/10.3389/fpubh.2023.1110777; https://www.ncbi.nlm.nih.gov/books/NBK585039/

229   https://pdfs.semanticscholar.org/c397/6ab6e6bb7eb6d8a6d631161a2e84302b6f03.pdf; https://www.ncbi.nlm.nih.gov/pmc/articles/PMC3184751/

230   https://doi.org/10.2174%2F1874210601206010143

231   https://www.nih.gov/news-events/nih-research-matters/how-stress-causes-hair-loss

232   https://pubmed.ncbi.nlm.nih.gov/25842469/

233   https://www.karger.com/article/fulltext/509001; https://www.ncbi.nlm.nih.gov/pmc/articles/PMC8906269/

234   https://www.frontiersin.org/articles/10.3389/fphar.2021.642980/full

INDEX

crooked teeth, 307–308

crowns and caps, 309

crow's feet, 46, 286–288, 296

curcumin, 131, 140, 343

D

dairy products, 70, 72, 114–115, 188

dark chocolate, 90–91, 94

dark circles under eyes, 61, 288–289

dark spots, 156, 174, 275, 342

deep breathing, 250–251

dehydration, 74–75

dental issues

crooked teeth, 307–308

crowns and caps, 309

discolorations, 308

teeth whitening, 310–312

toothpaste, 306–307, 310–311

deoxycholic acid, 281

dermabrasion, 299

dermaplaning, 267

dermatitis, 71, 147, 149, 165

dessert recipes, 221–224

diet and nutrition. *See also* fasting; supplements; *specific foods*

accelerated aging and, 52, 69, 74, 97–119, 333–334

antioxidant foods, 81, 84, 88–96

calorie restriction, 42–43, 46, 49, 53, 123–125

for collagen production, 52, 73, 78–86, 88

glycation and, 47, 52, 98–102, 106, 112, 117

gut microbes and, 70–74

importance of, 62–63

for inflammation reduction, 52, 69–77, 99, 185

intermittent fasting, 49, 53, 120–122, 125–131, 186–188, 196

in jump-starting autojuvenation, 183–195

ketogenic diets, 127, 129–130

macronutrients and micronutrients, 63–65

meal plans and recipes, 65–66

nutrient-dense foods, 49, 52, 65, 79, 102

premature aging and, 52, 62, 99–100, 111

processed foods, 48, 67, 79, 98, 102, 109–113, 118, 188

recipes, 196–234

for wrinkle reduction, 52, 84–85, 88, 173

dinner recipes, 209–217

discolored teeth, 308

DIY home treatments, 263–270

chemical peels, 264, 265

dermaplaning, 267

facials, 264

laser treatments, 265–266

microcurrent devices, 270

red light therapy, 266–267, 287

double chins, 280–281

droopy eyebrows, 290

duck lips, 305

monounsaturated fatty acids, 70, 76, 131

mood, assessment of, 36–37

multivitamins, 138–140, 189

muscle fibers, 252–253

muscles, assessment of, 36

N

naps, 240–241

nasolabial folds, 299, 301–302

neuroplasticity, 258

niacin (vitamin B$_3$), 64, 139

niacinamide, 170, 173, 274, 288

Nutrafol, 315, 319

nutrient-dense foods, 49, 52, 65, 79, 102

nutrition. *See* diet and nutrition

nuts, 52, 70, 82–83, 98, 108, 131

O

Obagi, Zein, 148

ochronosis, 169, 170, 275

octinoxate, 159

omega-3 fatty acids, 52, 69, 73, 75, 81–82, 99, 109–110, 131, 140, 343

omega-6 fatty acids, 69, 80, 81, 99, 109–110

oxidation, 47–48, 88, 112, 140, 342

oxidative stress, 48, 51, 53, 88, 118, 185

oxybenzone, 158–159

P

pantothenic acid (vitamin B$_5$), 64, 139

parabens, 171

PCOS (polycystic ovary syndrome), 102

PEG (polyethylene glycol), 172

peptides, 160, 167–168, 177–178, 189, 342

permafrown, 297

Petrucci, Kellyann, 80

phenol peels, 293

phthalates, 172

pigmentation, 288

platelet-rich plasma (PRP), 294–295, 318

Pollan, Michael, 112

pollution, 47, 53, 55, 67, 87–88, 151, 156, 176

polycystic ovary syndrome (PCOS), 102

polyethylene glycol (PEG), 172

polyphenols, 52, 91, 93, 131, 187

porcelain veneers, 308–309

positive thinking, 257

posture, assessment of, 35–36

power naps, 240–241

prebiotics, 52, 71, 73, 113

pregnancy, 72, 128, 166–167, 170, 307

premature aging

blood sugar spikes and, 101

diet and, 52, 62, 99–100, 111

free radicals and, 48

lifestyle factors and, 17

sun exposure and, 155–157

supplements for, 55

symptoms of, 33, 34, 34–37

wrinkles and, 33, 110

preservatives, 106, 112